DIGITAL PROCESSING OF SIGNALS

DIGITAL PROCESSING OF SIGNALS

BERNARD GOLD

and

CHARLES M. RADER

Lincoln Laboratory
Massachusetts Institute of Technology

with chapters by

ALAN V. OPPENHEIM

Research Laboratory of Electronics
Massachusetts Institute of Technology

and

THOMAS G. STOCKHAM, JR.

Computer Science Department
University of Utah

McGRAW-HILL
BOOK COMPANY

New York
St. Louis
San Francisco
London
Sydney
Toronto
Mexico
Panama

DIGITAL PROCESSING OF SIGNALS

Library of Congress Catalog Card Number 69-13606

ISBN 07-023642-9

11 12 13 14   K P K P   7 9 8 7 6 5 4

To Joe Levin

Preface

Electronic hardware steadily increases in complexity per unit cost and volume. Because of the importance of the computer, much effort has gone into the development of digital electronic components which are small, cheap, and fast. As this digital revolution proceeds, many processes previously performed exclusively by analog hardware, such as linear filtering and spectrum analysis, can be practically realized with digital hardware.

The purpose of this book is to present the theory, with some applications, of digital-signal-processing operations, with emphasis on the frequency-domain description of digital filtering and discrete spectrum analysis. Much of the material was used in a graduate course taught in the electrical engineering department at the Massachusetts Institute of Technology in 1965 and also as a summer course in 1967 at MIT. For the practicing engineer the book should be a useful reference in connection with computer simulations for signal-processing problems (in fields such as speech, seismology, medical research, oceanography, and radar), the design of special-purpose digital hardware, and the planning of general-purpose computer structures oriented toward signal processing.

The field of digital signal processing is too new to allow us to predict subsequent developments. Much of the material is based on our own research and that of the contributors, Alan V. Oppenheim and Thomas G. Stockham, Jr. Not enough scholarly attention has been given to the many published contributions; however, it appears that the present volume is the first that can claim some sort of comprehensive coverage of a field of growing importance.

The first chapter of the book introduces the two most significant digital-signal-processing operations: digital filtering and the discrete Fourier transform, emphasizing how the computation is performed by the digital hardware. Chapter 2 covers the basic theory of linear discrete systems and introduces and develops the z-transform notation used throughout the book. The focus in Chapter 3 is on digital-filter design in the frequency domain. Philosophically, and in many technical details, this chapter draws heavily from classical filter theory; however, it is important to realize that this classical filter theory is based on principles that transcend the hardware realization and can readily be applied to digital as well as analog systems. Chapter 4 deals with the digital-filter-synthesis problem, taking into account the finite register length in any digital computer.   Since many of the important issues in this chapter can be phrased in terms of random-noise models, some linear noise theory is included.   Chapter 5 briefly describes next-state simulation programming techniques.   Chapter 6 covers the theory of the discrete Fourier transform and includes a detailed explanation of the many possible forms of the fast Fourier transform.   In Chapter 7, the fast Fourier transform is applied to the realization of filtering and correlation operations.   In Chapter 8, the concept of generalized linear filtering is introduced and applied, via the discrete Fourier transform, to the problem of separating convolved signals.

We wish to pay tribute to our many talented coworkers in this new, engrossing field, whose published research and personal communications with us greatly helped us to write this book.   We wish especially to acknowledge the rewarding collaboration with Dr. M. J. Levin of the Lincoln Laboratory, which was cut short by his tragic death in an automobile accident.   We are grateful to Dr. J. F. Kaiser of the Bell Telephone Laboratories, who enthusiastically communicated many of his ideas to us, made us aware of new advances in the field, and critically and carefully reviewed our manuscript.

BERNARD GOLD
CHARLES M. RADER

Contents

PREFACE   vii

Chapter 1   A Discussion of Signal Processing   1

    1.1   Introduction   1
    1.2   Related Material   2
    1.3   Basic Computational Algorithms for Digital Filters   3
    1.4   Spectrum Analysis   6
    1.5   Realization of Digital-signal-processing Algorithms   7
    1.6   Further Remarks on Computer Facilities   8
    1.7   The Role of the Computer Display in Digital Signal Processing   9
    1.8   SUMMARY   13
          REFERENCES   18

Chapter 2   Discrete Linear Systems   19

    2.1   Model of a Discrete Linear System   19
    2.2   First-order Linear Difference Equation   21
    2.3   Frequency Response of First-order Network   24
    2.4   Geometric Interpretation of Frequency Response   24
    2.5 z-Transform   26
    2.6   Inverse z-Transform   27
    2.7   The Convolution Theorem   28
    2.8   The Complex Convolution Theorem   29
    2.9   Solution of First-order Difference Equation by z-Transform   30
   2.10   Solution of Second-order Difference Equation by z-Transform   31
   2.11   The Two-sided z-Transform   33
   2.12   Networks for Second-order Difference Equation   36
   2.13   Solution of mth-order Difference Equation by z-Transform
          Techniques   39
   2.14   Representation and Geometric Interpretation of System
          Function   42
   2.15   Discrete Networks with Nonzero Initial Conditions   43
   2.16   Other Realization of mth-order Systems   44
   2.17   SUMMARY   46
          PROBLEMS   46
          REFERENCES   47

ix

Chapter 3 Frequency-domain Design of Digital Filters   48

    3.1   Introduction   48
    3.2   Synthesis of Digital Filters   50
    3.3   Discussion of Several Digital-filter-design Techniques   50
    3.4   Technique 1:   Impulse Invariance   51
    3.5   Design of a Digital Lerner Filter That Is Impulse-invariant   52
    3.6   Gain of Digital Resonators   55
    3.7   Design of Digital Filters from Continuous Filters That Have
          Zeros at Infinity   56
    3.8   Review of Butterworth, Chebyshev, and Elliptic Filters   58
    3.9   Technique 2:   Digital-filter Specification from Squared-
          magnitude Function   66
    3.10 Technique 3:   Design of Digital Filters Using Bilinear
          Transformation of Continuous-filter Function   70
    3.11 Frequency-sampling Filters   78
    3.12 Technique 4:   Frequency-sampling Technique   81
    3.13 Design Examples   86
    3.14 SUMMARY   92
          APPENDIX   93
          REFERENCES   97

Chapter 4 Quantization Effects in Digital Filters   98

    4.1   General Discussion   98
    4.2   Errors Caused by Inexact Values of Constant Parameters   101
    4.3   Errors Caused by Analog-digital Conversion   103
    4.4   Analysis of Noise in First-order Filter Caused by A-D
          Converter   105
    4.5   Analysis of Digital Resonator with Two Poles and No Zeros   106
    4.6   Analysis of Digital Resonator with Two Poles and One Zero   107
    4.7   Errors Caused by Quantization of Products   107
    4.8   Deadband Effect   112
    4.9   Roundoff-noise Formulas for Different Arrangements of
          Digital Networks   116
    4.10 Example: Different Arrangement of Two-pole One-zero
          Network   117
    4.11 Noise Measurements   120
    4.12 Estimation of the Mean Value of the Noise   121
    4.13 The Autocovariance and Spectral Density   122
    4.14 Estimation of the Covariance Function and Spectral Density   123
    4.15 Example: Measurement of Noise in a Resonator Cascade   127
    4.16 SUMMARY   128
          REFERENCES   129

Chapter 5   Block-diagram Representation of Computer Programs   131

   5.1   Introduction   131
   5.2   The Object System   133
   5.3   Comparison of Source and Object Systems   136
   5.4   Elements for Object Systems   141
   5.5   Ordering of Execution of Subroutines in a Next-state
        Simulation   149
   5.6   The PATSI Compiler   152
   5.7   SUMMARY   158
        REFERENCES   158

Chapter 6   Discrete Fourier Transforms   159

   6.1   Continuous Fourier-transform Theory   159
   6.2   The Discrete Fourier Transform   162
   6.3   The Goertzel Algorithm   171
   6.4   The Fast Fourier Transform   173
   6.5   The Relation Between Decimation in Time and Decimation
        in Frequency   187
   6.6   Application of the Fast-Fourier-transform Algorithms   196
   6.7   SUMMARY   199
        REFERENCES   201

Chapter 7   High-speed Convolution and Correlation with Applications
        to Digital Filtering   *by Thomas G. Stockham, Jr.*   203

   7.1   Introduction   203
   7.2   The Product of DFTs: Periodic Convolution   204
   7.3   Aperiodic Convolution   205
   7.4   Partial Results and Sectioned Results   208
   7.5   Pairing in Real Convolution   211
   7.6   Choosing the Section Length   212
   7.7   Effects of Large N 212
   7.8   Discrete Fourier Transforms Expressed as Convolutions or
        Correlations   213
   7.9   High-speed Digital-filter Design   217
   7.10 SUMMARY   231
        REFERENCES   232

Chapter 8   Generalized Linear Filtering   *by Alan V. Oppenheim*   233

   8.1   Introduction   233
   8.2   Generalized Superposition   235

8.3   Generalized Linear Filtering   239
8.4   Homomorphic Filtering of Multiplied Signals   240
8.5   Homomorphic Filtering of Convolved Signals   242
8.6   Properties of the Complex Cepstrum   245
8.7   Minimum-phase Sequences and the Hilbert Transform   247
8.8   Sequences with Rational z-Transforms   250
8.9   Computation of the Complex Cepstrum   251
8.10   Applications of Homomorphic Filtering   253
8.11   Homomorphic Processing of Images   253
8.12   Homomorphic Processing of Speech   258
8.13   SUMMARY   262
       REFERENCES   263

INDEX   265

1
A Discussion
of Signal Processing

1.1 INTRODUCTION

Linear filtering and spectrum analysis are fundamental signal-processing operations of great utility in many branches of science and technology. These operations may be performed numerically (digitally) by special- or general-purpose digital computers as well as continuously by analog computers or RLC networks. Of special interest in many cases is the frequency-domain description of signals and linear systems. Such descriptions are valid for both continuous and discrete processing. The frequency-domain viewpoint is often preferable to the time-domain viewpoint when dealing with the processes of low-pass or bandpass filtering, differentiating, interpolating, and smoothing. In areas such as speech communications, seismology, sonar, radar, and medical technology, research workers often make use of these signal-processing techniques.

A growing body of examples exists where it has become preferable to perform digital rather than analog signal processing. This book is directed at the programmer who wants to simulate dynamical systems on

1

a general-purpose digital computer, at the designer who wants to build a special-purpose digital signal-processing device, and at the general-purpose-computer designer who wants to modify the basic computer structure so as to make programming of signal-processing operations more efficient. In short, it is a book devoted primarily to the theory of linear digital filtering and discrete spectrum analysis, with emphasis on the frequency-domain description of signals and systems.

1.2 RELATED MATERIAL

Historically, we can go back at least to Laplace [13]† who understood the z-transform which is the mathematical basis of numerical signal-processing operations. Kaiser [11], in his excellent survey, traces digital filtering back to the early 1600s and as being motivated by problems of astronomy and the compilation of mathematical tables. Several important treatises [2, 8, 14] have been published on the mathematics of linear difference equations; these can serve as good background material to the mathematically inclined in the same way that a book on linear differential equations is useful background for the study of Laplace transforms as applied to linear continuous systems. Of even more direct interest is the chapter by Hurewicz [7], the first modern work on sampled-data control systems. Interestingly, this chapter also lays the foundation for a good deal of the frequency-domain description of digital systems. Many of the subsequent books on sampled-data systems [4, 9, 10, 15, 16, 17, 20, 22] contain information which is most pertinent and makes for excellent supplementary reading to the present volume. To give several examples: Chapter 4 of Ragazzini and Franklin [17] treats discrete linear systems by z-transform analysis, and Chapter 10 treats the behavior of these systems when excited by random inputs. Chapters 1, 2, and 4 of Jury [10] contain a development of z-transform theory and its application to linear difference equations. Especially worth perusing for their insights into digital computation are the books by Blackman [1] and Hamming [6].

In addition to the above literature on discrete systems, the reader would be well rewarded to study some of the modern material on continuous linear systems. Of particular help in understanding Chap. 3 of this volume are the works of Guillemin [5], Storer [18], and Weinberg [21].

There are numerous contemporary papers on the subject of digital computation, digital filtering in the frequency domain, fast Fourier transform, high-speed convolution and correlation, and generalized linear filtering. These papers will be referred to at appropriate places in the book. An important aim of this book is to make the reader aware of and better able to understand this body of work.

† Bracketed numbers are keyed to the References at the end of the chapter.

1.3 BASIC COMPUTATIONAL ALGORITHMS FOR DIGITAL FILTERS

We have emphasized digital filtering and discrete Fourier transform (DFT) as the two basic signal-processing computational algorithms. This is not to say that they are the only algorithms of interest; for example, in Chap. 5 several nonlinear processes are mentioned, and in Chap. 8 great importance is attached to the logarithm. For purposes of illustrating the practical engineering problems that arise in the planning of a digital signal-processing system, we shall, in this section, confine our examples to filtering and DFT.

Let us imagine that a given digital filter has been specified in terms of its frequency-domain description and we want either to write a program or to design special-purpose digital logic for performing the required computational algorithm. First, it must be understood that there are many possible algorithms leading to theoretically identical filters.† Here we wish to distinguish between three broad classes of such algorithms, which we refer to as (1) convolution, (2) recursion, and (3) Fourier transformation. Depending on the particular filter, one of these three realizations will be the most desirable. We shall now define in somewhat heuristic terms the nature of these algorithms.

Consider a *signal* $x(nT)$. For simplicity in the present discussion, we constrain $x(nT)$ to be zero for $n < 0$, so that the signal consists of a sequence of numbers $x(0)$, $x(T)$, $x(2T)$, etc. The constant T is the sampling interval, so that implicit in our choice of notation is the possibility that $x(nT)$ was obtained by sampling a continuous signal $x(t)$ at uniformly spaced instants of time 0, T, $2T$, etc. In real-time applications, and for convenience in simulating systems that eventually must operate in real time, the explicit inclusion of T in the formulation is desirable. In the smoothing of data wherein physical time is not pertinent (for example, two-dimensional spatial filtering of a photograph), T can be set to unity, so that the signal is strictly a function of the index.

A linear digital filter can now be defined via the principle of super-position, as follows:

$$y(nT) = \sum_{m=0}^{n} h(mT)x(nT - mT) = \sum_{m=0}^{n} h(nT - mT)x(mT) \qquad (1.1)$$

Numerically, the output sequence $y(nT)$ is a weighted sum over all previous values of the input sequence. The weights $h(nT)$ define the filter. As can be seen from (1.1), if the input is the particular sequence 1, 0, 0, 0, . . . , the output sequence is precisely $h(nT)$. In this book, we shall

† This remark is true if the lengths of all registers containing the numerical signals and parameters are sufficient to ensure that quantization effects can be neglected. In Chap. 4 these effects are described in some detail.

be interested only in stable digital filters; hence if the input sequence is bounded, the output sequence must also be bounded. It can be shown that a necessary and sufficient condition for stability [7] is

$$\sum_{n=0}^{\infty} |h(nT)| < \infty \qquad (1.2)$$

Inequality (1.2) is satisfied if the unit-pulse response $h(nT)$ is truncated, that is, if $h(nT) = 0$ except for $N_1 < n < N_2$. It can also be satisfied by unit-pulse responses of unlimited duration, for example, $h(nT) = K^n$ with $|K| < 1$. In the latter case, the computational algorithm defined by (1.1) eventually becomes unfeasible, since an increasing number of computations are necessary, as n increases, for each subsequent value of the output sequence. Since (1.2) implies that $h(nT)$, even of theoretically infinite duration, tends to zero for large enough n, it is possible to compute an approximate output sequence by simply truncating $h(nT)$. This appears to be too empirical an approach to be generally satisfactory; as a result, a body of theory based on filters with truncated impulse response has developed. The less duration that can be assigned to the impulse response, the more efficient will be the computation indicated in (1.1). Kaiser [11] has reviewed many useful techniques of frequency-domain digital-filter design, using filters with truncated impulse response.

A second fundamental computational algorithm is based on the theory of linear difference equations with constant coefficients. A simple example is the equation

$$y(nT) = Ky(nT - T) + x(nT) \qquad y(-T) = 0 \qquad (1.3)$$

where K is a constant less than unity. The algorithm is realized by computing each successive $y(nT)$ from the previous computation $y(nT - T)$ and the most recently arrived signal sample $x(nT)$. More generally, programs can be written or hardware can be built for performing computations of systems of linear difference equations, each of the form

$$y(nT) = \sum_{k=1}^{m} K_k y(nT - kT) + \sum_{k=0}^{r} L_k x(nT - kT) \qquad (1.4)$$

Equations (1.3) and (1.4) define computational algorithms that generate the solution $y(nT)$ for successive values of n. Such equations can be theoretically analyzed as linear dynamic systems, and, in particular, the response of these systems to sampled sinusoidal inputs can be derived in terms of the coefficients L_k and K_k.

The theory presented in Chaps. 2 and 3 enables the designer to deduce the set of difference equations that yield the desired frequency response and, in particular, to determine how complex an algorithm is needed, since, in general, the more severe the requirements on the filter (sharper cutoff, greater adherence to some ideal characteristic, etc.), the more coefficients are needed in the difference equations. Since the choice of difference equations is, in general, not unique, many design options are open; further choice is possible with the help of the considerations of Chap. 4, which relate the synthesis problem to quantization effects. In that chapter it will be shown that memory register lengths may often be shortened by a wise choice of the specific filter configuration.

The third fundamental algorithm is based on the discrete Fourier transform (DFT). How a DFT can be used to yield a computation equivalent to (1.1) is not obvious and will be discussed in some detail in Chap. 7. Briefly, the difficulty arises because the product of two DFTs in the frequency domain is equivalent to *circular* convolution in the time domain whereas (1.1) is a linear convolution; this difficulty can be overcome by an appropriate sectioning technique as described in Chap. 7. Neglecting until then this aspect of the computation, we shall describe this DFT algorithm for digital-filter synthesis.

The DFT of a signal $x(nT)$ is defined as

$$X(k\Omega) = \sum_{n=0}^{N-1} x(nT)e^{-j\Omega Tnk} \qquad k = 0, 1, 2, \ldots, N-1 \qquad (1.5)$$

In this equation, $j = \sqrt{-1}$ and $\Omega = 2\pi/NT$, where N is the number of signal samples to be transformed. To show how (1.5) can be used for a digital-filter algorithm, we quote the convolution theorem, proved in Chap. 6: If $X(k\Omega)$ is the DFT of $x(nT)$ and $H(k\Omega)$ is the DFT of $h(nT)$, then the product $H(k\Omega)X(k\Omega)$ is the DFT of the convolution of $x(nT)$ and $h(nT)$:

$$H(k\Omega)X(k\Omega) = \mathrm{DFT}\left\{ \sum_{n=0}^{N-1} x(nT)h(((m-n))T) \right\} \qquad (1.6)$$

where $((m-n))$ signifies $(m-n)$ modulo N. If it is assumed that this circular convolution can be equivalent to a linear convolution, the computational procedure can now be outlined.

1. Compute the DFT of the signal $x(nT)$. Call this $X(k\Omega)$.
2. Multiply $X(k\Omega)$ by $H(k\Omega)$, where $H(k\Omega)$ is the desired frequency response of the filter at the frequencies $2\pi k/NT$, $k = 0, 1, 2, \ldots, N-1$.
3. Compute the inverse DFT of the product $H(k\Omega)X(k\Omega)$. This yields the desired output:

$$y(nT) = \frac{1}{N} \sum_{k=0}^{N-1} X(k\Omega)H(k\Omega)e^{j\Omega Tnk} \qquad (1.7)$$

In almost all cases, the practical application of this algorithm makes use of the fast Fourier transform (FFT). Suitably used, this technique can be employed to synthesize filters with either finite-length or infinite-length impulse response.† In other words, any filter realizable by the computations (1.1) or (1.4) can also be synthesized via the DFT algorithm. In cases where the length of the filter impulse response is fairly long, say of a length of several hundred samples, this technique is almost always computationally more efficient than the direct convolution (1.1).

We have tried to make clear in the above discussion that the designer of digital filters has available a rather rich storehouse of design and synthesis techniques. The primary aim of Chaps. 2, 3, 4, 6, and 7 is to present the theoretical foundations of these techniques.

1.4 SPECTRUM ANALYSIS

Discrete spectrum analysis of an input signal can be performed either with digital filters or via the DFT. A filter-bank spectrum analyzer could be composed of a set of parallel channels, as shown in Fig. 1.1, where the energy at a given spectral location is measured by a bandpass digital filter tuned to the proper frequency, followed by a square-law device and a low-pass digital filter to smooth the power measurement. The designer of a spectrum analyzer has many parameters to choose: the number of input-data points to be analyzed, the frequency resolution, the list of frequencies where spectral energy is to be measured, the weighting function that mul-

† In Chap. 7, the synthesis of finite-impulse-response filters by FFT will be treated in detail. At this writing, it is not commonly understood that the FFT can be used to synthesize filters with infinite-length impulse response. The practical implications of this result are not yet clear, and it will not be treated further in the present volume.

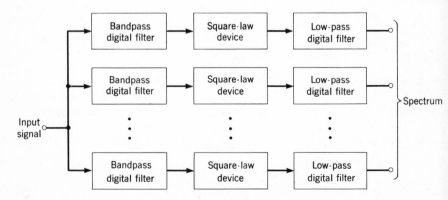

Fig. 1.1 Digital spectrum analyzer.

tiplies the signal, the time intervals between successive spectral measurements, and perhaps how to average these successive measurements. Clearly, these choices are so completely dependent on the problem at hand that no general design statements can be made. The point to be stressed is that the computational algorithms described in Sec. 1.3 are also applicable to the realization of digital spectrum analyzers. Cooley, Lewis, and Welch [3] have written a good treatise on many of the issues of spectrum analysis, using the FFT algorithm.

1.5 REALIZATION OF DIGITAL SIGNAL-PROCESSING ALGORITHMS

A central issue in contrasting digital with analog signal-processing systems is this: Each analog filter in an aggregate of filters requires a complete realization, distinct from all other filters. Different digital filters can share a common arithmetic element, although they must have distinct memory elements. For example, assume that seismic data, sampled 20 times per second, is being processed. With present-day circuit speeds, the time required to compute a new output point for even a fairly sophisticated filter might be, say, 50 μsec. Thus, up to 1,000 such filter algorithms could be computed in a single sampling time of 50 msec, so that a single arithmetic unit could service a bank of 1,000 filters. Perhaps more to the point is the fact that the extra cost of having 1,000 rather than 100 filters would derive almost exclusively from the increased memory requirements, little or no extra cost being added to the price of control and arithmetic. We deduce from this argument the general thought that digital rather than analog processing becomes preferable as the signal-processing facility or device increases in complexity.

Let us now consider a specific example, to clarify the influence of present-day component technology on the planning of a digital signal processor. We should preface this example with the remark that, whereas theoretical developments have a relatively long life span, hardware-component specifications change much more rapidly.

Our example is that of a general-purpose digital-computer program to simulate a 30-channel spectrum analyzer of a speech waveform. As we shall see in Chap. 2, the linear difference equation for a digital resonator involves two or three multiply operations and several additions; most present-day scientific computers could perform the computation in about 10 instructions, which could be executed in about 40 μsec. To simulate the 30 resonators requires 1,200 μsec of execution time. Thus real-time spectrum analysis of speech is not possible on these general-purpose computers, since a speech wave must be sampled about once each 100 μsec. This example gives an idea of the increase in speeds needed to simulate speech communication systems in real time. Similar arguments can be used to show that the above-mentioned computers cannot simulate, in real time, many geophysics, medical, and radar processing systems.

However, the rapid advances presently taking place in the manufacture of digital components tempt us to predict that operations such as the simulation of a real-time speech-spectrum analyzer will soon become feasible by digital means. Integrated circuit memories with read cycles of 150 nsec are now available commercially. Basic operations such as data transfers and addition can be performed in less than 100 nsec. Speeds of thin-film and core memories continue to increase above 1-MHz rates. Multiplication of two 18-bit numbers can be performed in less than 1 μsec. Computer facilities an order of magnitude faster than those mentioned above appear to be possible.

In addition to faster components, changes in the logical organization of general-purpose computers could make possible large additional speed increases. What is needed is some form of parallel processing, but how to do this generally and yet economically is not yet clear. In fact, this book will have accomplished part of its purpose if it helps to clarify the issues for the computer-logic designer faced with the problem of designing a digital signal-processing facility.

1.6 FURTHER REMARKS ON COMPUTER FACILITIES

A computer must have other, more nebulous qualities to make it suitable for signal processing. One such quality is the degree of intimacy between research worker and computer. "Batch" processing, whereby the programmer hands the computer operator a card deck and returns for the printout based on the program run, is an example of little intimacy. The other extreme is exemplified by much of the work done on the Lincoln Laboratory TX-2 computer, where the programmer, who is also the only man working on a given project, spends several hours alone with the computer. In this situation, the abilities of man and machine to communicate rapidly become very important. A great aid to such communication is the set of peripheral devices attached to the computer; of these, the computer-controlled oscilloscope is most useful. Analog-to-digital and digital-to-analog converters, pulse interrupts, an accurate external timing source, a multiplexer and demultiplexer, knobs, and toggle switches are further examples of useful computer-controlled external devices other than the standard items such as keyboards, tape and card readers, and punches and high-speed printers; all these devices help to make on-line computer operation more meaningful.

In our opinion, a high degree of intimacy between research worker and computer is of great psychological benefit. How to bring this about economically is still a controversial issue. Certainly, more intimacy is obtained by increasing processing speeds. A large time-shared computer with many geographically separated keyboards provides a limited on-line facility economically. However, this is generally not a signal-processing facility, since no direct signal inputs or outputs are permissible nor is a display oscilloscope available. To convert an individual time-shared console into a signal-processing facility not only requires an expensive console but greatly enlarges the hardware and software requirements of the central processor.†

An alternative approach involves the individual small- or medium-sized computer, complete with the desired external equipment. With such a facility serving a working group of about 10 scientific programmers, an individual will have full use of the computer about as often as he needs it for prolonged periods. The disadvantage is the relatively small amount of high-speed memory available in a modest facility. Despite this, the present trend among signal-processing workers is toward a small or medium "private" facility. A possible compromise between the above directions is the use of modest facilities that are loosely coupled to a large central processor in such a way that the small computers can be used alone but can also take advantage of the large memory store and expensive peripheral devices associated with the large computer.

.7 THE ROLE OF THE COMPUTER DISPLAY IN DIGITAL SIGNAL PROCESSING

The research worker must often be able to hear or see the results of the processing. For example, after simulation of a speech communication system, one wants to hear the processed speech. In programs for automatically extracting the fundamental frequency of a speech wave it is desirable visually to display the speech concurrently with the results of the processing. Seismic waves, cardiograms, brain waves, and radar returns are profitably studied visually. Thus it is logical to attach a computer-controlled oscilloscopic display to a signal-processing computer facility.

It is outside the scope of this book to treat problems in computer graphics. There are many good references on this subject [19]. We restrict the following brief discussion to display examples that have been found useful for speech communication simulation.

On the most obvious level, a computer display can portray the signal in a rather versatile way. Programs are easily written to display any number of lines, and compression and expansion of the time scale are easily programmed. Figures 1.2 and 1.3 show the same speech wave.

Figure 1.4 shows a speech wave where the words being spoken

† This remark is based on observations of the Lincoln Laboratory TX-2 computer time-sharing system.

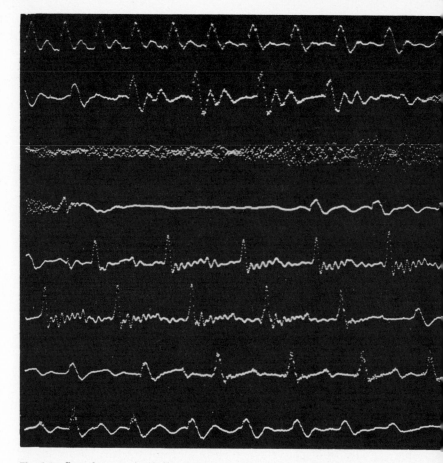

Fig. 1.2   Speech wave (each line follows the line above it in time).

are synchronously displayed.   Also displayed are automatically derived
period markers.   Building special-purpose laboratory hardware to achieve
this is rather formidable.   To perform this task in a reasonable time even
with a computer display is nontrivial.   It depends on the ability of a
listener, on hearing the audio tape run at half speed, to tap a key in rhythm
with the spoken syllable, so that a two-track audio tape containing the
original speech and reasonably synchronous timing markers can be pre-
pared.   Once this is done, a computer with real-time analog-to-digital
(A-D) input and pulse-interrupt capability can create Fig. 1.4.   Time-
varying spectral displays, being, in essence, three-dimensional, are much
more difficult to design, and for these a computer display has extreme
advantages over attempts to build special-purpose hardware.   The reason

s that the ultimate efficiency of the display can be judged only by eye so
that much trial-and-error testing is necessary.   Figures 1.5 to 1.7 show
several ways of displaying the speech spectrum available from a filter-
bank spectrum analyzer; the horizontal axis is time, and the vertical axis
is frequency, in discrete steps.

Displays are often useful in determining whether a given program is
working correctly and performing as desired.   Figure 1.8 shows the frequency response of a programmed digital filter.   Generating such a display is, of course, straightforward once the required computations have
been made.   The advantage of the display (as opposed to a printout) is
twofold: (1) Gross errors in the computation program are more quickly
detected and (2) any desired empirical adjustments on the filter param-

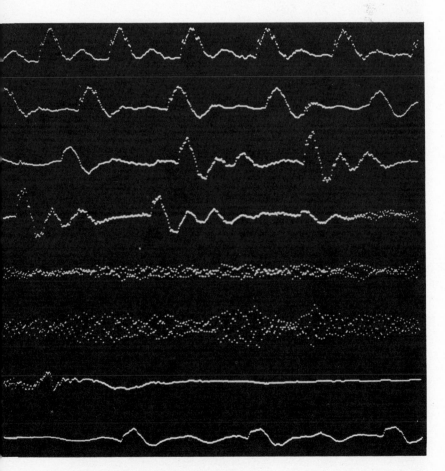

Fig. 1.3   Speech wave with expanded time scale.

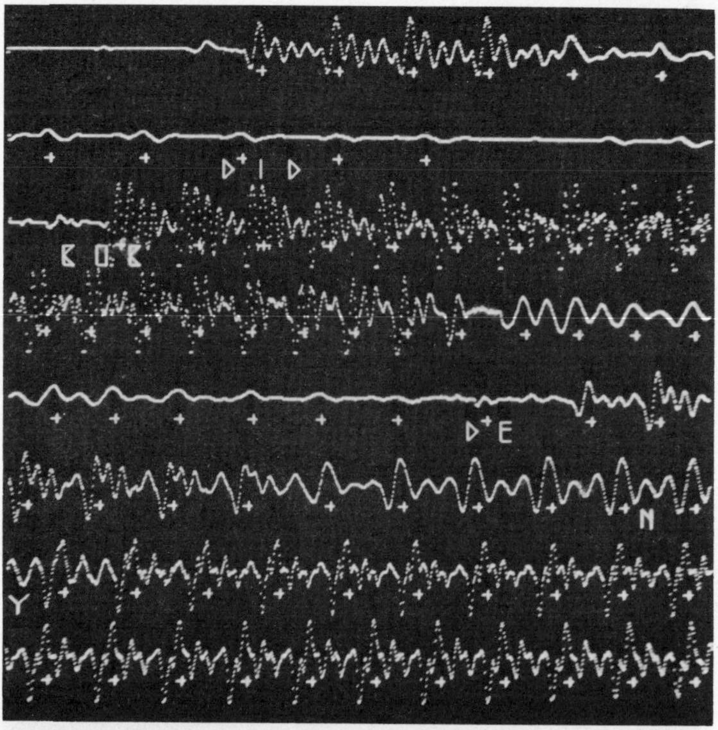
D I D
B O B
D E
N
Y

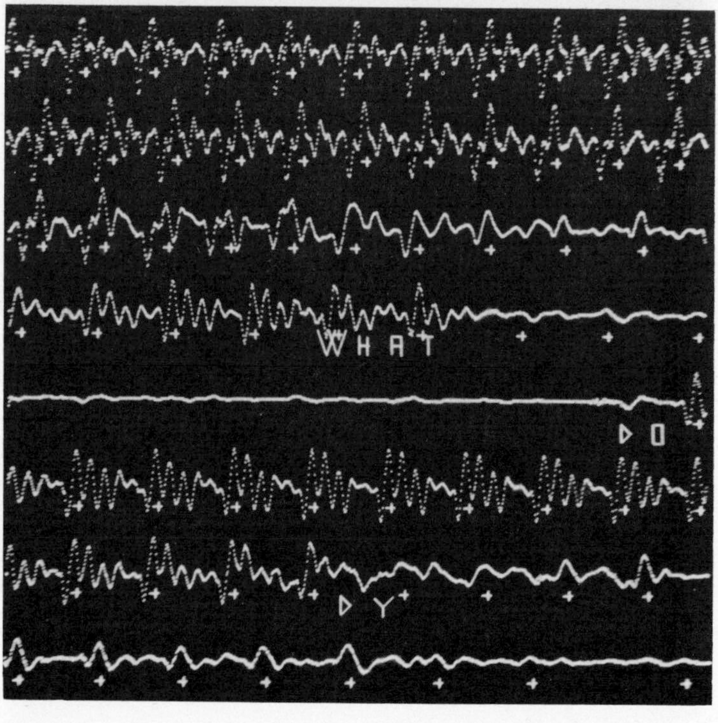
W H A T
D O
D Y

eters are more quickly made. Figure 1.9 shows how the side lobes of the filter were empirically reduced by on-line adjustment of several parameters.

1.8 SUMMARY

Chapter 1 is in itself primarily a summary of the remainder of the book. It began with a heuristic discussion of the linear filtering operation. An important point is that all three mathematical formulations of the digital-filter problem, namely, convolution, recursion, and the z-transform, have practical computational counterparts. Also worth noting is that recursive techniques and the FFT are in many cases competitive ways of performing the same operations and that the correct choice is often not obvious. The present revolution in digital-hardware development makes it likely that real-time digital systems using filters and spectrum analyzers may supersede many existing analog systems; this requires a good understanding of the theory presented in this book.

The chapter closed with a brief section on displays. In the context

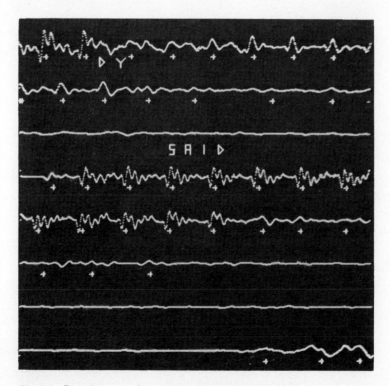

Fig. 1.4   Speech wave with synchronized words (see opposite page also).

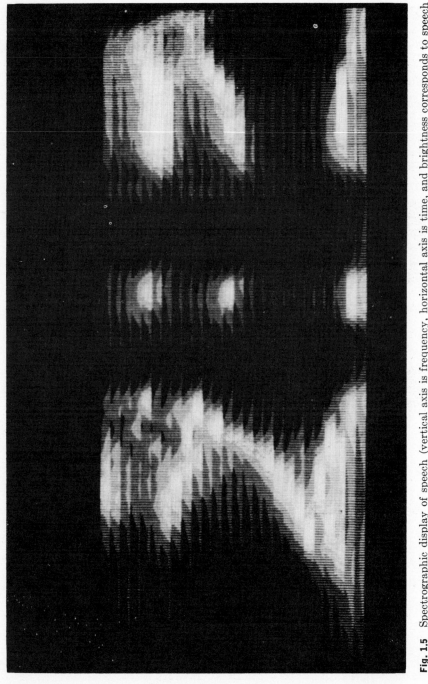

Fig. 1.5 Spectrographic display of speech (vertical axis is frequency, horizontal axis is time, and brightness corresponds to speech intensity).

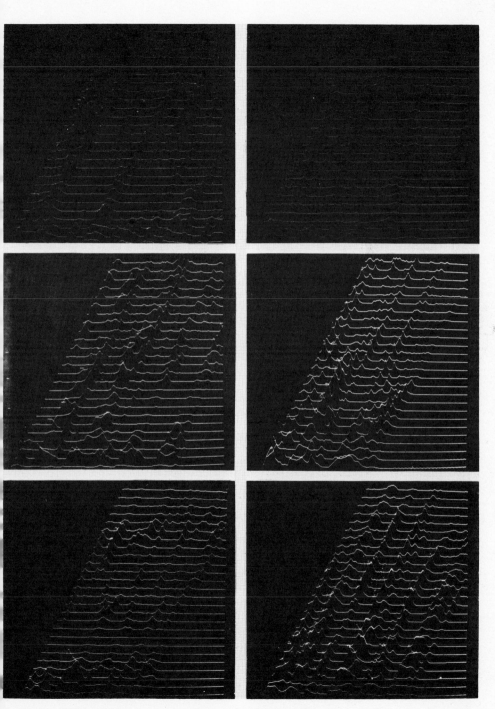

Fig. 1.6 Display of spectra of speech from filter-bank analyzer.

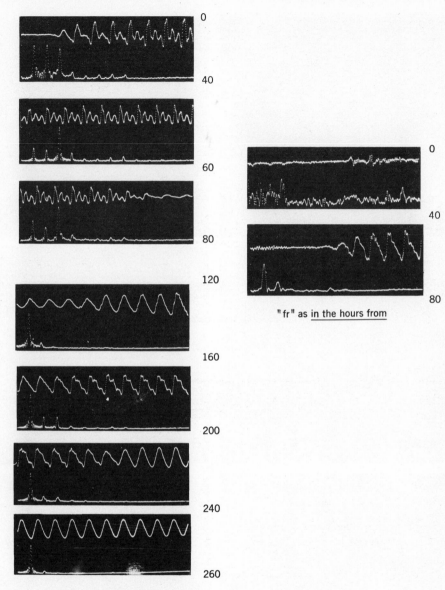

"hour" as in the hours from

"fr" as in the hours from

Fig. 1.7   Segments of speech and corresponding spectra as measured by discrete Fourier transform.

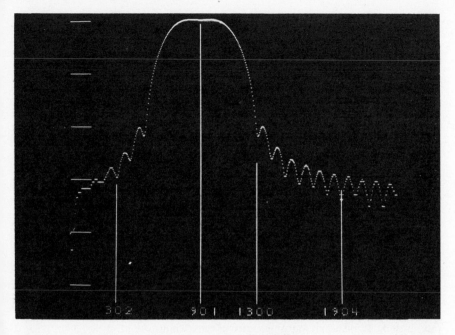

Fig. 1.8 Amplitude vs. frequency response of digital filter.

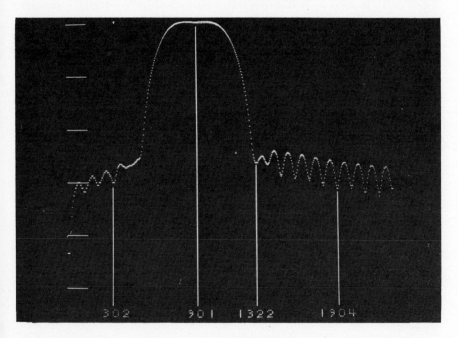

Fig. 1.9 Amplitude vs. frequency response of digital filter with reduced side lobes.

of our material, we have little to say about theoretical aspects of displays but we encourage workers in all fields of signal processing to develop strong intuitions on using oscilloscopic displays.

REFERENCES

1. Blackman, R. B.: "Linear Data-smoothing and Prediction in Theory and Practice," Addison-Wesley Publishing Company, Inc., Reading, Mass., 1965.
2. Boole, G.: "Finite Differences," out of print.
3. Cooley, J. W., P. Lewis, and P. Welch: The Fast Fourier Transform and Its Applications, *IBM Res. Paper RC*-1743, Feb. 9, 1967.
4. Freeman, H.: "Discrete-time Systems," John Wiley & Sons, Inc., New York, 1965.
5. Guillemin, E. A.: "Synthesis of Passive Networks," John Wiley & Sons, Inc., New York, 1957.
6. Hamming, R. W.: "Numerical Methods for Scientists and Engineers," McGraw-Hill Book Company, New York, 1962.
7. James, H. M., N. B. Nichols, and R. S. Phillips: "Theory of Servomechanisms," pp. 231–261, McGraw-Hill Book Company, New York, 1947.
8. Jordan, C.: "Calculus of Finite Differences," Chelsea Publishing Company, New York, 1960. (Reprint of 1939 edition.)
9. Jury, E. I.: "Sampled-data Control Systems," John Wiley & Sons, Inc., New York, 1958.
10. Jury, E. I.: "Theory and Application of the z-Transform Method," John Wiley & Sons, Inc., New York, 1964.
11. Kuo, F. K., and J. F. Kaiser (eds.): "System Analysis by Digital Computer," chap. 7, John Wiley & Sons, Inc., New York, 1967.
12. Lanczos, C.: "Applied Analysis," Prentice-Hall, Inc., Englewood Cliffs, N. J., 1956.
13. Laplace, P. S.: "Ouevres Complètes," 1779.
14. Milne-Thomson, L. M.: "The Calculus of Finite Differences," Macmillan & Co., Ltd., London, 1933.
15. Mishkin, E., and L. Braun, Jr.: "Adaptive Control Systems," pp. 119–183, McGraw-Hill Book Company, New York, 1961.
16. Monroe, A. J.: "Digital Processes for Sampled Data Systems," John Wiley & Sons, Inc., New York, 1962.
17. Ragazzini, J. R., and G. F. Franklin: "Sampled-data Control Systems," McGraw-Hill Book Company, New York, 1958.
18. Storer, J. E.: "Passive Network Synthesis," McGraw-Hill Book Company, New York, 1957.
19. Sutherland, I. E.: Sketchpad: A Man-Machine Graphical Communication System, *Proc. Spring Joint Computer Conf.*, 1963.
20. Tou, J. T.: "Digital and Sampled-data Control Systems," McGraw-Hill Book Company, New York, 1959.
21. Weinberg, L.: "Network Analysis and Synthesis," McGraw-Hill Book Company, New York, 1962.
22. Wilts, C. H.: "Principles of Feedback Control," pp. 195–220, Addison-Wesley Publishing Company, Inc., Reading, Mass., 1960.

2
Discrete Linear Systems

2.1 MODEL OF A DISCRETE LINEAR SYSTEM

In Sec. 1.3 we described heuristically the different algorithms for the synthesis of linear digital systems. This chapter deals primarily with the theory of the linear difference-equation algorithms as exemplified by (1.3) and (1.4). These equations, which are often called *recursive* and *autoregressive*, serve as the starting point for the various modes of description of discrete linear systems, namely, via frequency response, digital networks, geometrical interpretation in the complex z plane, and z-transform operational notation. Much of the material for this chapter can be found in the references cited in Chap. 1. In addition, the reader might find the papers by Kaiser [3] and Rader and Gold [4] to be useful background material. Since the intent of this chapter is primarily to familiarize the reader with the technical tools needed to comprehend much of the remainder of this book, a problem set is included to help him achieve this end.

Linear (analog) network theory is based on the electrical properties of inductances, capacitances, and resistances which lead, via Kirchhoff's

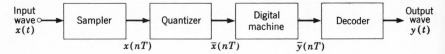

Fig. 2.1 Model of digital machine for processing analog signals.

laws, to a description of networks in terms of linear differential equations with constant coefficients. By contrast, discrete or digital linear-systems theory begins with sets of linear difference equations with constant coefficients, which can then be realized by manipulating numbers in a special or general-purpose computer. To realize the algorithm for a difference equation, the essential factor is that the input signal be composed of discretely spaced samples, usually in time. All considerations in this book are based on uniformly spaced samples. Nonuniform spacing of samples is treated in the literature [2] but is outside the scope of this book.

A large number of pertinent problems can be modeled as shown in Fig. 2.1. These are the class of problems wherein the signal to be processed is an analog signal (for example, the output of a microphone or seismometer) but where the processing is to be performed digitally. Not all problems fall in this category; for example, a digital speech synthesizer need have no analog inputs. In Fig. 2.1, the analog signal is sampled at time intervals T to produce the sequence $x(nT)$. The effects of this sampling process have been treated extensively; nearly all the references given in Chap. 1 cover this subject, which is described by either of the terms *folding* or *aliasing*. In frequency-domain terms, this means that the spectrum of the analog signal is repeated periodically at frequencies $1/T$, $2/T, 3/T$, etc. Thus, if the sampling rate is lower than the Nyquist rate of the signal (that is, if T is greater than the inverse of twice the signal bandwidth), the spectrum of the sampled signal, so to speak, folds back on itself, creating a frequency distortion and making impossible the reconstitution of the analog signal by linear filtering.

The quantizer of Fig. 2.1 is a necessary part of any actual digital processor. It should be stressed that quantization effects occur also within the digital machine of Fig. 2.1. These effects occur, for example, in specifying the constant coefficient of a linear difference equation by means of a number in a finite-length register; such an effect can be considered static in that the properties of the difference equation are changed once and for all. There are also dynamic effects when signals are multiplied by either coefficients or other signals and the product is either rounded off or truncated to a given finite-length register. In this chapter and in Chap. 3 we make the important assumption that these errors can be neglected. In Chap. 4 we shall study these effects in some detail and show how they can influence the synthesis of digital filters.

In Fig. 2.1, if a continuous output is desired, the output samples $\bar{y}(nT)$ of the digital machine are passed through a decoder or desampler, which creates a continuous signal from the pulse train. Most commonly, the decoder is a digital-to-analog converter followed by a linear analog filter which removes the aliased frequencies caused by the sampling process. Thus, for example, if $T = 10^{-4}$ sec, the decoder would be a low-pass filter with cutoff frequency 5 kHz. An especially simple case of such a filter is a sample-and-hold circuit.

The study of continuous linear dynamic systems is greatly facilitated by the introduction of the operational methods of Laplace and Fourier transforms and also by the use of network concepts. Equivalently, the study of linear discrete systems benefits from the introduction of the z-transform and the use of network concepts. To readers familiar with continuous linear systems, many of the theoretical developments for discrete linear systems will seem most reminiscent, and indeed they should seem so. But it is well to remember that there is a discrete, or digital, framework within which new insights must be developed, and attempts to deduce results in this framework from already digested information from the analog domain are likely to cause more rather than less distress.

2.2 FIRST-ORDER LINEAR DIFFERENCE EQUATION

As an example of a simple first-order linear difference equation, consider

$$y(nT) = Ky(nT - T) + x(nT) \tag{2.1}$$

In Fig. 2.1, $x(nT)$ represents the sampled input signal. Removing the quantizer causes $\bar{x}(nT)$ to be equal to $x(nT)$ so that $\bar{x}(nT)$ can be eliminated from consideration. The digital machine in Fig. 2.1 accepts the sequence $x(nT)$; each time a new input arrives, the computation indicated in (2.1) is performed and the output $y(nT)$ is either presented to the decoder or saved or processed further within the digital machine. It is easy to visualize the structure of a machine that performs the operations required by (2.1). Three storage registers are needed, one for $x(nT)$, one for $y(nT)$, and one for K. The product $Ky(nT - T)$ is obtained, the contents of the register containing $x(nT)$ are added to this product, and the result is stored in the register containing $y(nT)$. The digital machine is now ready for the next input sample.

At the instant before a new input sample is applied (and the algorithm commenced) the state of the system is completely defined by the number in the register containing $y(nT)$. Without loss of generality, we can call this number $y(-T)$ and assume that $x(0)$ is about to be applied to the digital machine. From (2.1) we see that $y(0) = Ky(-T) + x(0)$, $y(T) = K^2y(-T) + Kx(0) + x(T)$, and by induction we can arrive at

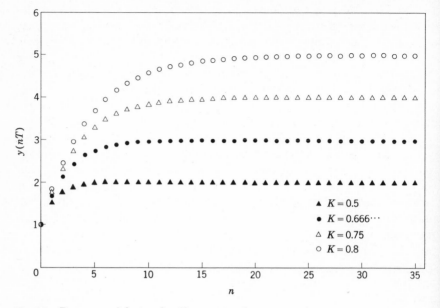

Fig. 2.2   Response of first-order filter to step input.

the general result:

$$y(nT) = K^{n+1}y(-T) + \sum_{i=0}^{n} K^i x(nT - iT) \tag{2.2}$$

The solution to (2.1) is thus given explicitly by (2.2).

If the input $x(t)$ in Fig. 2.1 is a unit step function, the sequence $x(nT)$ is unity for $n = 0, 1, 2, \ldots$ In this case, (2.2) becomes

$$y(nT) = K^{n+1}y(-T) + \frac{1 - K^{n+1}}{1 - K} \tag{2.3}$$

For large values of n, a steady-state solution $y(nT) = 1/(1 - K)$ is approached. However, this solution is valid only for $|K| < 1$; otherwise, we can see from (2.2) that $y(nT)$ increases without limit.

Figure 2.2 shows $y(nT)$ versus n for several values of K, with $y(-T) = 0$. We see a resemblance between these curves and the RC step-response curve shown in Fig. 2.3. Making $K \geq 1$ is akin to introducing a negative resistance in the analog RC circuit. We shall observe that many similarities exist between analog and digital networks. However, they are in no sense equivalent, only similar.

A convenient picture of (2.1) is shown in Fig. 2.4. This can be related to the computation algorithm as follows: When the new data sam-

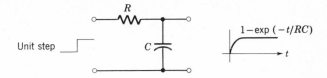

Fig. 2.3 Response of RC circuit to step input.

ple $x(nT)$ appears, the previously computed value of the output $y(nT)$ becomes the $y(nT - T)$ for the new iteration. The multiplication and addition are performed, resulting in the updating of the output register to $y(nT)$.

We call Fig. 2.4 a digital, or discrete, network, or, if there can be no misunderstanding, simply a network. In this book, many such networks will be discussed and analyzed, and so it is worth making a few remarks on notation. The rectangular figure with z^{-1} inside will be used to denote a unit delay of duration equal to the sampling interval T. The circle with a plus sign denotes addition of all the inputs (arrows pointing into circle), with a single output. Multiplication by a fixed coefficient is represented by writing the coefficient alongside the signal path.

It is of interest to inquire what the actual hardware realization of (2.1) would be. Let us define several registers: K to hold the multiplication constant K, Y to hold $y(nT)$, and X to hold $x(nT)$. Then, when a new sample $x(nT)$ is entered in X, the operation $Y \times K + X \to Y$ takes place and a single iteration is accomplished.

Thus we see that three registers are needed, as well as addition and multiplication operations. Clearly, the realization of (2.1) is more expensive than a comparable passive analog network unless, perhaps, the network is to operate at very low frequencies, where the capacitor size becomes large. However, if it is desired to realize a number of networks rather than a single one, only two registers per network need be added, registers equivalent to K and Y. Thus, for example, if a set of 500 networks has to be realized and if the arithmetic unit speeds are sufficiently high, all networks could use the same adder and multiplier. To illustrate, let us assume a sampling rate of 100 cps. Then 500 difference equations

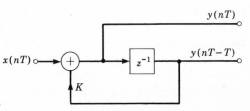

Fig. 2.4 Digital first-order network.

must be executed in 10 msec, so that the above set of operations must each be done in less than 20 μsec to achieve a real-time system.

2.3 FREQUENCY RESPONSE OF FIRST-ORDER NETWORK

A most valuable description of our simple system is in terms of its response to a sinusoidal input. If $x(t)$ in Fig. 2.1 is taken to be $e^{j\omega t}$, then $x(nT) = e^{jn\omega T}$. The response to a real sine wave, say $x(t) = \cos \omega t$, can always be found by adding the responses to inputs $\frac{1}{2}e^{j\omega t}$ and $\frac{1}{2}e^{-j\omega t}$. Now, the solution to (2.1) for an input $x(nT) = e^{jn\omega T}$ can be found by straightforward induction. Here we give only the steady solution with initial condition $y(-T) = 0$. This is

$$y(nT) = \frac{e^{jn\omega T}}{1 - Ke^{-j\omega T}} = \frac{x(nT)e^{j\omega T}}{e^{j\omega T} - K} \qquad (2.4)$$

As we expect, the output $y(nT)$ is a complex exponential, just like the input, but modified by a *transfer function*, which we can define as

$$H(e^{j\omega T}) = \frac{e^{j\omega T}}{e^{j\omega T} - K} = |H|e^{j\psi} \qquad (2.5)$$

with

$$
\begin{aligned}
|H| &= \frac{1}{(1 + K^2 - 2K \cos \omega T)^{\frac{1}{2}}} \\
\psi &= \omega T - \tan^{-1} \frac{\sin \omega T}{\cos \omega T - K}
\end{aligned}
\qquad (2.6)
$$

Equations (2.6) introduce the idea of frequency selectivity. In Fig. 2.5 the magnitude of H from (2.6) is plotted for various values of K and $T = 1$, from which we can detect both the similarities and differences between the digital system and a corresponding simple analog system such as an RC circuit, with $K = e^{-T/RC}$.

2.4 GEOMETRIC INTERPRETATION OF FREQUENCY RESPONSE

Equations (2.5) and (2.6) can be interpreted geometrically by reference to the unit circle shown in Fig. 2.6.

We see that $|H|$ in (2.5) is $1/d$ and $\psi = \omega T - \delta$, so that the vector from any point on the circle (at angle ωT) to the critical point K on the real axis completely defines the transfer function. We shall see that this geometric picture can be generalized to describe linear digital networks having an arbitrary number of operations.

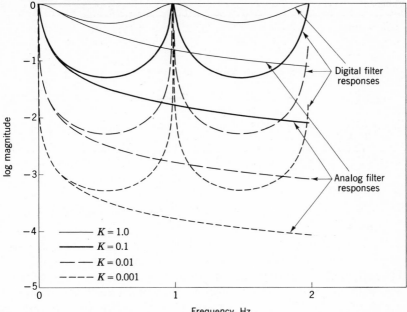

Fig. 2.5 Comparison of digital and analog first-order network frequency-response functions.

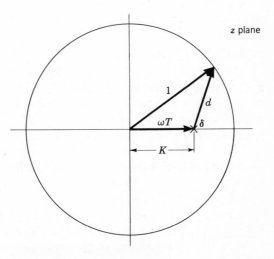

Fig. 2.6 z-plane representation of first-order network.

Table 2.1

Description of sequence	Sequence	z-transform of sequence
Unit pulse	$x(nT) = 1$ for $n = 0$ $= 0$ for $|n| > 0$	$X(z) = 1$
Unit step	$x(nT) = 1$ for $n \geq 0$ $= 0$ for $n < 0$	$X(z) = \dfrac{1}{1 - z^{-1}}$
Complex exponential	$x(nT) = e^{in\omega T}$ for $n \geq 0$ $= 0$ for $n < 0$	$X(z) = \dfrac{1}{1 - e^{i\omega T}z^{-1}}$

2.5 z-TRANSFORM

We have seen that a single first-order difference equation can be visualized in terms of a transfer function which shows how the system behaves as a function of frequency when excited by a sinusoidal forcing function. Also, we have been able to picture this transfer function geometrically. The formal tool for extending this interpretation to the general linear difference equation is the z-transform, which permits algebraic manipulation of difference equations much as the Laplace transform does for differential equations. We shall briefly examine the properties of the z-transform, next discuss linear difference equations from a networks viewpoint, and then apply the z-transform methods to find general solutions of the equations.

Consider a sequence of numbers $x(0)$, $x(T)$, $x(2T)$, . . . , $x(nT)$, . . . , which presumably were derived by sampling a continuous waveform $x(t)$. The z-transform of this sequence is defined to be

$$X(z) = \sum_{n=0}^{\infty} x(nT)z^{-n} \qquad (2.7)$$

z is a complex variable and $X(z)$ is a function of a complex variable. Since (2.7) is a power series in the variable z^{-1}, the question of convergence of this series naturally arises. Hurewicz [1] has treated this problem in some detail, and we shall quote a few basic results from his work.

The series (2.7) converges for $|z| > R$ and diverges for $|z| < R$, where the radius of convergence R is the upper limit of the sequence $|x(nT)|^{1/n}$, $n = 1, 2, 3, ...$ Thus, for example, if $x(nT) = K^n$, (2.7) converges outside a circle of radius K.

For $|z| > R$, $X(z)$ is an analytic function of z. Thus the function defined by (2.7) and extended by the process of analytic continuation can be called the z-transform of the sequence $x(nT)$. In this way, $X(z)$ is defined over the entire z plane but the expression (2.7) has validity only in the region of convergence.

2.6 INVERSE z-TRANSFORM

By definition, $x(nT)$ is the inverse z-transform of $X(z)$. By manipulation of (2.7) and the Cauchy integral theorem, $x(nT)$ can be found explicitly. First, multiply both sides of (2.7) by z^{k-1} and then perform a closed-line integration on both sides of the equation. If the integration path is within the regions of convergence of the infinite series of (2.7), summation and integration can be interchanged, yielding

$$\oint X(z)z^{k-1}\,dz = \sum_{n=0}^{\infty} x(nT) \oint z^{k-n-1}\,dz \qquad (2.8)$$

The Cauchy theorem tells us that if the path of integration encloses the origin then $\oint z^{k-n-1}\,dz = 0$ unless $k = n$. For $k = n$ the integral becomes $2\pi j$. This, combined with (2.8), results in the z-transform inversion theorem.

$$x(kT) = \frac{1}{2\pi j} \oint X(z)z^{k-1}\,dz \qquad (2.9)$$

Let $x(nT) = K^n$. Then $X(z) = 1/(1 - Kz^{-1}) = z/(z - K)$. To verify that $x(nT)$ is the inverse z-transform of $X(z)$, we use (2.9), performing the integral around a circle of radius larger than K. This yields

$$x(nT) = \frac{1}{2\pi j} \oint \frac{z^n\,dz}{z - K} \qquad (2.10)$$

Equation (2.10) is solved by the residue theorem, yielding $x(nT) = K^n$ if the path of integration encloses the pole at $z = K$. Thus a suitable path is the circle C_1 of radius $K + \epsilon$ shown in Fig. 2.7, where ϵ can be made

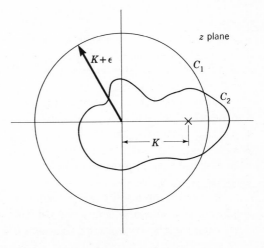

Fig. 2.7 Acceptable integration paths for inverse z-transform.

arbitrarily small. However, the path C_2 and any number of other paths enclosing the pole will also work for this case.

If $K > 1$, then, according to the discussion in Sec. 2.5, the region of convergence lies outside the unit circle in the complex z plane. In most cases, the sequence $x(nT) = K^n$ is not of physical interest for $K > 1$ since the sequence grows unboundedly as n increases, and such sequences can be classified as unstable. Thus the unit circle would be the smallest circle in the z plane that would fall inside the region of convergence for all stable sequences of the form K^n. This property of the unit circle can be extended to all stable sequences, and this accounts for the widespread use of the unit circle as the contour used for the inverse z-transform.

2.7 THE CONVOLUTION THEOREM

Let $X(z)$ be the z-transform of $x(nT)$ and $H(z)$ be the z-transform of $h(nT)$. Then, if $Y(z) = X(z)H(z)$ is the z-transform of $y(nT)$, we shall show that

$$y(nT) = \sum_{m=0}^{n} x(mT)h(nT - mT) = \sum_{m=0}^{n} x(nT - mT)h(mT)$$

$$(2.11)$$

A simple way to prove this relationship is by induction, through examination of the product:

$$X(z)H(z) = [x(0) + x(T)z^{-1} + x(2T)z^{-2} + \cdots]$$
$$\times [h(0) + h(T)z^{-1} + h(2T)z^{-2} + \cdots]$$
$$= x(0)h(0) + z^{-1}[x(0)h(T) + x(T)h(0)]$$
$$+ z^{-2}[x(0)h(2T) + x(T)h(T) + x(2T)h(0)] + \cdots$$
$$= y(0) + z^{-1}y(T) + z^{-2}y(2T) + z^{-3}y(3T) + \cdots \quad (2.12)$$

The reader will readily verify, by equating coefficients of equal powers of z^{-1}, that (2.11) is valid.

The convolution theorem, as in the analog case, can be taken as the defining equation of linear discrete systems. Figure 2.8 illustrates the computational aspect of (2.11). The dashed curve represents $h(nT - mT)$, and the solid curve represents $x(mT)$. The sum, up to n, of the term-by-term products of the two sequences is thus the convolution sum, as given by the middle expression of (2.11).

If $h(nT)$ represents the sequence created by a linear discrete network when a unit pulse (first entry in Table 2.1) is applied to the network, (2.11) defines the response of this network to the arbitrary input $x(nT)$.

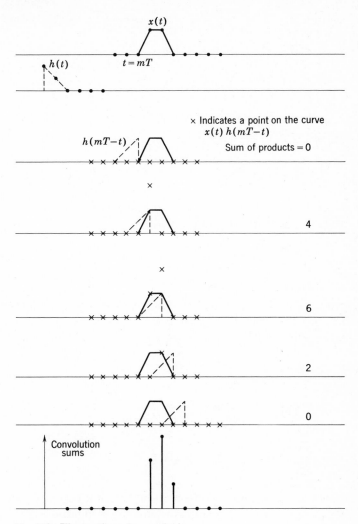

Fig. 2.8 Illustration of convolution.

2.8 THE COMPLEX CONVOLUTION THEOREM

In Sec. 2.7 we showed that the product of two z-transforms corresponds to the convolution of the sequences. In this section we shall derive the z-transform of the product of two sequences. Let

$$U(z) = \sum_{n=0}^{\infty} x(nT)y(nT)z^{-n} \qquad (2.13)$$

with $X(z)$ the z-transform of $x(nT)$ and $Y(z)$ the z-transform of $y(nT)$.

Therefore,

$$y(nT) = \frac{1}{2\pi j} \oint Y(v)v^{n-1}\, dv$$

$$x(nT) = \frac{1}{2\pi j} \oint X(v)v^{n-1}\, dv$$

(2.14)

In accordance with Sec. 2.6, we choose the integration contours to be the unit circle. Thus,

$$U(z) = \sum_{n=0}^{\infty} x(nT)z^{-n} \frac{1}{2\pi j} \oint Y(v)v^{n-1}\, dv$$

Interchanging integration and summation and identifying the resulting summation as a z-transform lead to

$$U(z) = \frac{1}{2\pi j} \oint Y(v)X\left(\frac{z}{v}\right) v^{-1}\, dv$$

(2.15)

Equation (2.15) is sometimes called the complex convolution theorem. That it indeed has the form of a convolution can be demonstrated by recalling that the path of integration is the unit circle, so that, if we make the substitutions $v = e^{j\theta}$, $z = re^{j\varphi}$, we obtain

$$U(re^{j\varphi}) = \frac{1}{2\pi} \int_0^{2\pi} Y(e^{j\theta})X(re^{j(\varphi-\theta)})\, d\theta$$

(2.16)

An interesting special case of (2.15) is the case $x(nT) = y(nT)$ and $z = 1$. From (2.13) and (2.14), we derive

$$\sum_{n=0}^{\infty} y^2(nT) = \frac{1}{2\pi j} \oint Y(v)Y\left(\frac{1}{v}\right) v^{-1}\, dv$$

(2.17)

which allows expression of the mean-squared value of a signal in terms of its z-transform. These relations are of use in noise studies, as we shall see in Chap. 4. The perceptive reader may have noticed that, if $Y(v)$ is presumed to have poles *within* the unit circle, then $Y(1/v)$ has poles outside the unit circle, so that the question of integration must be carefully examined. Since this question is closely tied to the problem of two-sided z-transforms, we shall defer further discussion until the two-sided z-transform is treated.

2.9 SOLUTION OF FIRST–ORDER DIFFERENCE EQUATIONS BY z-TRANSFORM

Returning now to the first-order equation (2.1), we shall apply the z-transform to solve this equation. Multiplying both sides of (2.1) by z^{-n} and summing from zero to infinity, we obtain

$$\sum_{n=0}^{\infty} y(nT)z^{-n} = K \sum_{n=0}^{\infty} y(nT - T)z^{-n} + \sum_{n=0}^{\infty} x(nT)z^{-n} \qquad (2.18)$$

Recalling the definition, (2.7), of the z-transform and using the substitution $m = n - 1$ in the first term of the right side of (2.18), we obtain

$$Y(z) = Kz^{-1}Y(z) + Ky(-T) + X(z) \qquad (2.19)$$

or

$$Y(z) = \frac{X(z)}{1 - Kz^{-1}} + \frac{Ky(-T)}{1 - Kz^{-1}} \qquad (2.20)$$

An explicit solution for $y(nT)$ can now be found from the inversion theorem (2.9), provided that a closed-form expression for $X(z)$ was found from the given input sequence $x(nT)$. For example, if $x(nT) = e^{jn\omega T}$, then, from Table 2.1, $X(z) = 1/(1 - e^{j\omega T}z^{-1})$ and

$$y(nT) = \frac{1}{2\pi j} \oint \frac{z^{n+1}}{z - e^{+j\omega T}} \frac{dz}{z - K} + \frac{y(-T)}{2\pi j} \oint \frac{z^n \, dz}{z - K} \qquad (2.21)$$

where we assume that the path of integration is a circle bigger than the unit circle by an arbitrarily small amount.

The result of integrating (2.21) is

$$y(nT) = \frac{K^{n+1}}{K - e^{j\omega T}} + \frac{e^{j(n+1)\omega T}}{e^{j\omega T} - K} + y(-T)K^n \qquad (2.22)$$

For $K < 1$, the first and third terms of (2.22) decay exponentially with increasing n. The middle term is the steady-state response of the system to an exponential input and is seen to be identical to the solution (2.4).

2.10 SOLUTION OF SECOND-ORDER DIFFERENCE EQUATION BY z-TRANSFORM

Greater insight into the frequency selectivity of a discrete system is obtained by examining a second-order equation, where the phenomenon of resonance is encountered. Consider the equation

$$y(nT) = K_1 y(nT - T) + K_2 y(nT - 2T) + x(nT) \qquad (2.23)$$

with initial conditions $y(-T) = y(-2T) = 0$. Applying the z-transform,

$$Y(z) = \frac{X(z)}{1 - z^{-1}K_1 - z^{-2}K_2} = \frac{z^2 X(z)}{z^2 - K_1 z - K_2} = H(z)X(z) \qquad (2.24)$$

where $H(z)$ depends entirely on the network and thus can be interpreted as a transfer function. If we choose as the input the unit impulse (first entry in Table 2.1), then $X(z) = 1$, and $y(nT)$ may be found by the inverse z-transform

$$y(nT) = \frac{1}{2\pi j} \oint \frac{z^{n+1}\,dz}{(z - \gamma_1)(z - \gamma_2)} \tag{2.25}$$

where

$$\gamma_{1,2} = \frac{K_1}{2} \pm \sqrt{\frac{K_1^2}{4} + K_2}$$

from which we obtain

$$y(nT) = \frac{1}{\gamma_1 - \gamma_2}\,(\gamma_1^{n+1} - \gamma_2^{n+1}) \tag{2.26}$$

The most interesting case occurs for the condition $K_1^2/4 + K_2 < 0$, so that γ_1 and γ_2 become complex numbers $\gamma_1 = re^{jbT}$, $\gamma_2 = re^{-jbT}$ with $r^2 = -K_2$ and $2r \cos bT = K_1$. Substituting these into (2.25) yields

$$y(nT) = \frac{r^n}{\sin bT}\,\sin\,(n + 1)bT \tag{2.27}$$

Equation (2.27) is seen to represent a damped sinusoid decaying exponentially to zero but meanwhile oscillating at a natural radian frequency b. In fact, the result obtained is the sampled equivalent of the response of a simple RLC resonant circuit to an impulse. We should thus expect the frequency-selective properties of (2.23) to be like those of an RLC circuit. That this in indeed true can be ascertained by applying a sinusoidal oscillation $x(nT) = \cos n\omega_0 T$ in (2.23). It can then be shown that the steady-state portion of the output $y(nT)$ is given by

$$y(nT) = |H| \cos\,(n\omega_0 T + \psi)$$

where the complex function $|H|e^{j\psi}$ is exactly equal to $[H(z)]_{z=e^{j\omega_0 T}}$ as defined in (2.24) and is thus a frequency-dependent function. From (2.24) and utilizing the roots found in (2.25), we can write

$$H(e^{j\omega T}) = \frac{e^{2j\omega T}}{(e^{j\omega T} - \gamma_1)(e^{j\omega T} - \gamma_2)} \tag{2.28}$$

which can immediately be interpreted geometrically, as in Fig. 2.9, to be proportional to the *vector ratio* $u^2/d_1 d_2$. Notice that, as the point P moves close to the pole γ_1, the distance d_1 decreases and the value of $|H|$ increases. We see that a resonance is exhibited which is sharper as the poles move closer to the unit circle. Explicit expressions for $|H|$ and ψ

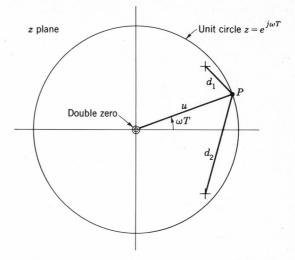

Fig. 2.9 z-plane representation of second-order digital filter.

can be found directly from the geometry of Fig. 2.9 and are

$$|H| = \left[\frac{1}{(1 - K_1 \cos \omega T - K_2 \cos 2\omega T)^2 + (K_1 \sin \omega T + K_2 \sin 2\omega T)^2} \right]^{\frac{1}{2}}$$

$$\psi = - \tan^{-1} \frac{K_1 \sin \omega T + K_2 \sin 2\omega T}{1 - K_1 \cos \omega T - K_2 \cos 2\omega T} \tag{2.29}$$

The value of $|H|$ is plotted in Fig. 2.10 for $T = 10^{-4}$, a resonant frequency of 1,500 Hz, and several values of r. The magnitude $|H|$ and phase ψ must be periodic in ωT; this follows from the original premise of a sampled input and is embodied in the unit-circle z-plane representation. It is also clear from (2.27) that the system is unstable for $r > 1$.

2.11 THE TWO-SIDED z-TRANSFORM

Thus far, discussion has been restricted to the "causal" discrete linear system, that is, a system that cannot respond before being excited by an input signal. This restriction has been implicit in the difference equations, wherein an output is always a function of previous outputs, never of future outputs. Once the concept of frequency response has been introduced, it is often an aid to thinking to remove the causal restriction. For example, a filter with a rectangular magnitude function and linear phase in the passband is a useful mathematical entity which can be inserted in a

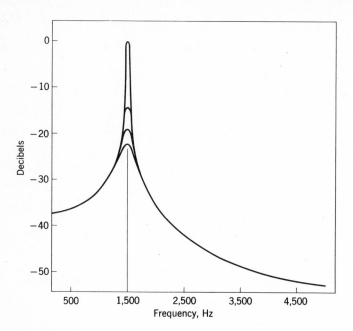

Fig. 2.10 Frequency response of digital resonator for several damping constants.

system, even though it is physically unrealizable. Such filters are easily described for discrete problems by using the two-sided z-transform.

Instead of the definition (2.7), let us define the two-sided z-transform:

$$X(z) \triangleq \sum_{n=-\infty}^{\infty} x(nT)z^{-n} \tag{2.30}$$

which can also be written

$$X(z) = \sum_{n=0}^{\infty} x(nT)z^{-n} + \sum_{n=0}^{\infty} x(-nT)z^{n} - x(0) \tag{2.31}$$

The first sum of (2.31) is the one-sided z-transform, which converges for $|z| > R_1$, where R_1 is the upper limit of the sequence $[x(nT)]^{1/n}$. The second converges for $|z| < R_2$, where R_2 is the upper limit of the sequence $[x(-nT)]^{-1/n}$. For example, if $x(nT) = K^{|n|}$, the two-sided z-transform has the ring of convergence shown in Fig. 2.11.

The inversion theorem can be derived following the procedure of Sec. 2.6 and leads to exactly the same formula as (2.9). Integration can be carried out around the unit circle as long as the convergence conditions specified in the preceding paragraph are satisfied.

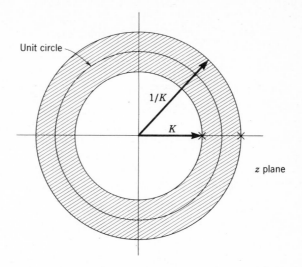

Fig. 2.11 Acceptable integration for inversion of two-sided z-transform.

EXAMPLE

Given a discrete linear filter with the frequency vs. magnitude function (and zero phase shift everywhere) shown in Fig. 2.12, find the weighting function of this filter. The weighting function is defined as the response to a unit pulse at $n = 0$.

By using (2.9) with $H(z) = X(z)$ and $h(kT) = x(kT)$ and letting $z = e^{j\omega T}$,

$$h(kT) = \frac{1}{2\pi} \int_{-\omega_c}^{\omega_c} e^{jk\omega T} \, d\omega = \frac{\sin k\omega_c T}{k\pi}$$

Notice that $h(kT)$ is an even function of kT. This is always true if the phase of the network is taken to be zero. If the filter has linear phase, it can be shown that the impulse response is as shown in Fig. 2.13 except for a time translation.

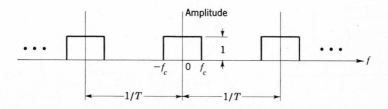

Fig. 2.12 "Ideal" rectangular frequency response of digital filter.

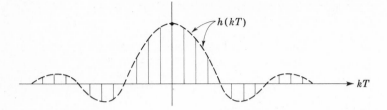

Fig. 2.13   Unit-pulse (or impulse) response of rectangular filter.

2.12   NETWORKS FOR SECOND-ORDER DIFFERENCE EQUATION

The second-order difference equation

$$y(nT) = K_1 y(nT - T) + K_2 y(nT - 2T) + x(nT) - Lx(nT - T) \qquad (2.32)$$

can be represented by Fig. 2.14.

The z-transform relationships for (2.32) are

$$Y(z) = \frac{X(z)(1 - Lz^{-1})}{1 - K_1 z^{-1} - K_2 z^{-2}} = H(z)X(z) \qquad (2.33)$$

The network $H(z)$ has a zero and two poles.   It can be seen from Fig. 2.14 that the zero is created by delay elements in the *forward* path, i.e., by direct delay of the input, whereas poles are created by delays of the feedback output.   In fact, an equivalent network $H(z)$ can be obtained by reversing the order of the feedforward and feedback delays, as shown in Fig. 2.15.

A third network with $H(z)$ equivalent to that of Figs. 2.14 and 2.15 is shown in Fig. 2.16.   This form is easily derived from Fig. 2.15 by noting that the two rightmost delays in Fig. 2.15 are redundant in that the same signal is fed into them.

The network of Fig. 2.16 is called *canonic* because the delays are used for both poles and zeros and thus this system has the minimum num-

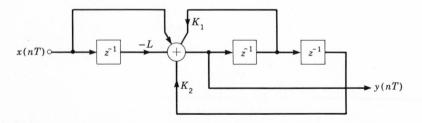

Fig. 2.14   Direct form for two-pole one-zero digital network.

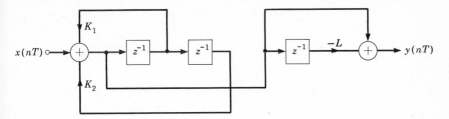

Fig. 2.15 Alternative form for two-pole one-zero network.

ber of delay elements for a given-order network, i.e., n *delays* for nth order. We shall see later that, although all the forms of Figs. 2.14 to 2.16 yield identical z-transforms, their behavior in the presence of noise caused by the roundoff error of the multiplications is different.

Yet another realization of the second-order system of two poles is shown in Fig. 2.17. This network is described by the equation pair

$$y_1(nT) = Ay_1(nT - T) + By_2(nT - T) + Lx(nT)$$
$$y_2(nT) = Cy_1(nT - T) + Dy_2(nT - T)$$

(2.34)

The z-transforms of both equations may be taken, and $Y_1(z)$ or $Y_2(z)$ may be solved explicitly, yielding

$$Y_1(z) = \frac{LX(z)(1 - Dz^{-1})}{1 - (A + D)z^{-1} + (AD - BC)z^{-2}}$$
$$Y_2(z) = \frac{LCX(z)}{1 - (A + D)z^{-1} + (AD - BC)z^{-2}}$$

(2.35)

It is easily found that $Y_1(z)$ and $Y_2(z)$ are networks with poles having radial distance $r = (AD - BC)^{1/2}$ from the center of the unit circle at the angles given by $\pm \cos^{-1}(A + D)/2r$. $Y_2(z)$ has a second-order zero at

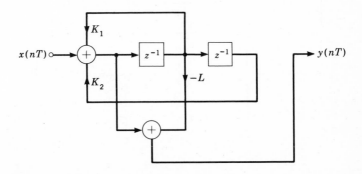

Fig. 2.16 Canonic form for two-pole one-zero network.

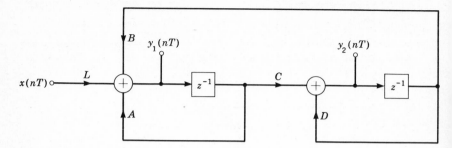

Fig. 2.17 Coupled form for two-pole network.

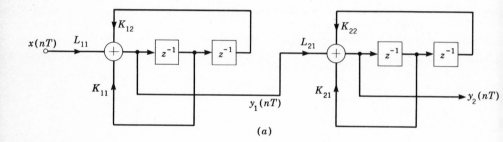

$$(a)$$

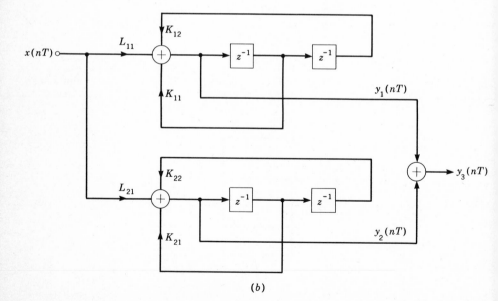

$$(b)$$

Fig. 2.18 (a) Serial or cascade form; (b) parallel form.

$z = 0$, and $Y_1(z)$ has two real-axis zeros at $z = D$ and $z = 0$. Thus the outputs $y_1(nT)$ and $y_2(nT)$ are exactly like those of the second-order systems previously described. This system appears more complex, since four multiplications rather than two or three are needed. But, as we shall see in Chap. 4, this extra freedom in the choice of parameters often proves useful in designing a filter so as to minimize the adverse effects of parameter quantization and can thus lead to a shorter word length of the hardware realization.

The networks we have described can be expanded into more complex systems by arranging several of them in serial, parallel, or mixed serial and parallel combinations. For example, the two equations

$$\begin{aligned} y_1(nT) &= K_{11}y_1(nT - T) + K_{12}y_1(nT - 2T) + L_{11}x(nT) \\ y_2(nT) &= K_{21}y_2(nT - T) + K_{22}y_2(nT - 2T) + L_{21}y_1(nT) \end{aligned} \quad (2.36)$$

constitute a serial arrangement whereby the output $y_1(nT)$ of the first equation is used as input to the second equation, as seen in Fig. 2.18a. Similarly, the equations

$$\begin{aligned} y_1(nT) &= K_{11}y_1(nT - T) + K_{12}y_1(nT - 2T) + L_{11}x(nT) \\ y_2(nT) &= K_{21}y_2(nT - T) + K_{22}y_2(nT - 2T) + L_{21}x(nT) \\ y_3(nT) &= y_1(nT) + y_2(nT) \end{aligned} \quad (2.37)$$

shown in Fig. 2.18b, constitute a parallel arrangement.

Digital networks in series are naturally isolated from one another, in distinction to continuous systems where rather sophisticated isolation amplifiers are required to prevent interaction between adjacent RLC networks. This explains to some extent why the emphasis in designing RLC networks is on the physical realization; this has resulted in a large body of elegant and fairly complex synthesis theory. This theory appears to be unnecessary for discrete-filter design.

2.13 SOLUTION OF mTH–ORDER DIFFERENCE EQUATION BY z-TRANSFORM TECHNIQUES

The general mth-order difference equation is given by

$$y(nT) = \sum_{i=0}^{r} L_i x(nT - iT) - \sum_{i=1}^{m} K_i y(nT - iT) \quad (2.38)$$

The form of (2.38) emphasizes the iterative nature of the difference equation; given the m previous output values $y(nT - T)$, $y(nT - 2T)$, etc., and the $r + 1$ most recent values of the input $x(nT)$, $x(nT - T)$, etc., the

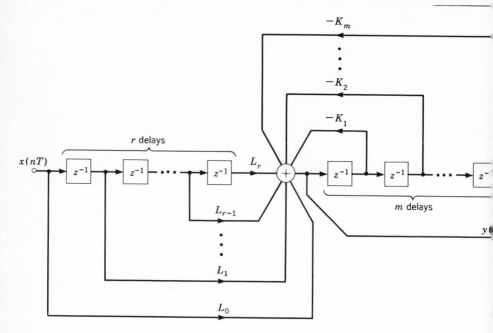

Fig. 2.19 Direct form for mth-order network.

new output may be computed.   The pictorial representation of (2.38) is given in Fig. 2.19.

The z-transform may now be used to obtain a general solution for $y(nT)$ in terms of the input and the network.   In this development, we assume the system to be initially at rest, so that all the x's and y's in (2.38) are zero when the iteration begins.   If we rewrite (2.38) as

$$\sum_{i=0}^{m} K_i y(nT - iT) = \sum_{i=0}^{r} L_i x(nT - iT) \qquad K_0 = 1 \tag{2.39}$$

and take the z-transform of both sides, we obtain

$$\sum_{i=0}^{m} K_i \sum_{n=0}^{\infty} y(nT - iT)z^{-n} = \sum_{i=0}^{r} L_i \sum_{n=0}^{\infty} x(nT - iT)z^{-n} \tag{2.40}$$

Since the z-transform of a sequence delayed by i samples is equal to the z-transform of the original sequence multiplied by z^{-i} if the initial values are zero, (2.40) becomes

$$Y(z) \sum_{i=0}^{m} K_i z^{-i} = X(z) \sum_{i=0}^{r} L_i z^{-i}$$

or

$$Y(z) = X(z) \frac{\sum\limits_{i=0}^{r} L_i z^{-i}}{\sum\limits_{i=0}^{m} K_i z^{-i}} = X(z)H(z) \qquad (2.41)$$

Thus $Y(z)$ is explicitly determined as the product of the input z-transform $X(z)$ and a system function $H(z)$ which is a rational fraction in z^{-1} and is a function of the constant coefficients in the original difference equation. By the inverse z-transform, the output $y(nT)$ is explicitly obtained.

If we choose as the input sequence

$$x(nT) = \begin{cases} 1 & \text{for } n = 0 \\ 0 & \text{for } n \neq 0 \end{cases}$$

then $X(z) = 1$. Clearly, the response to this input is the inverse z-transform of $H(z)$. Thus the sequence 1, 0, 0, 0, . . . plays the part in digital-network theory that the unit pulse plays in continuous-filter theory. We shall refer to $h(nT)$, the inverse z-transform of $H(z)$, as the impulse response of the digital filter $H(z)$.

From (2.41) we can also derive the canonic form of the mth-order difference equation. Defining an intermediate sequence $w(nT)$ with associated z-transform $W(z)$ such that

$$W(z) = \frac{X(z)}{\sum\limits_{i=0}^{m} K_i z^{-i}} \qquad (2.42)$$

$$Y(z) = W(z) \sum_{i=0}^{r} L_i z^{-i}$$

we derive from (2.42) the simultaneous difference equations

$$w(nT) = x(nT) - \sum_{i=1}^{m} K_i w(nT - iT)$$

$$y(nT) = \sum_{i=0}^{r} L_i w(nT - iT) \qquad (2.43)$$

which lead to the canonic structure shown in Fig. 2.20.

The primary importance of the system function $H(z)$ for the purposes of filter theory is in its interpretation as a frequency-selective function. Let us assume that the input is a sampled complex exponential wave.

$$x(nT) = e^{jn\omega T} \qquad (2.44)$$

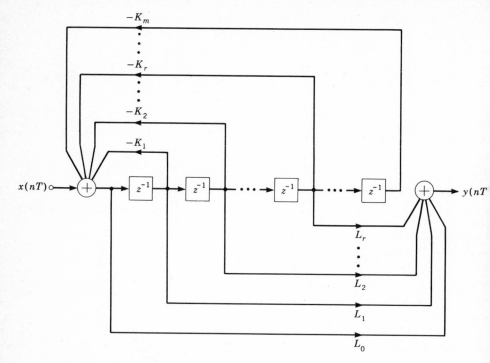

Fig. 2.20 Canonic form for mth-order network.

The solution for $y(nT)$ of (2.38) for such an input can be written

$$y(nT) = F(e^{j\omega T})e^{jn\omega T} \tag{2.45}$$

By substitution of (2.45) and (2.44) into (2.38), we quickly arrive at the result

$$F(e^{j\omega T}) = \frac{\displaystyle\sum_{i=0}^{r} L_i e^{-ji\omega T}}{\displaystyle\sum_{i=0}^{m} K_i e^{-ji\omega T}} = H(e^{j\omega T}) \tag{2.46}$$

2.14 REPRESENTATION AND GEOMETRIC INTERPRETATION OF SYSTEM FUNCTION

We see that the system function $H(z)$ of (2.41) is interpretable as a frequency-response function for values of z on the *unit circle* in the complex z plane. Note that the radian frequency ω is a continuous frequency so that the physical significance of the frequency-response function is the same as that for continuous systems. Furthermore, the system function is a rational fraction whose numerator and denominator can be fac-

ꓲored, so that $H(z)$ is uniquely defined, except for a constant multiplier, ꓐy the positions of its poles and zeros in the z plane, and its value for any ꓑoint z determined directly by the distances of that point from the singuꓲarities. We are thus led directly to a geometric interpretation whereby the value of the frequency-response function for any frequency ω is obtained by rotating by an angle ωT about the circle, measuring the distances to the zeros R_1, R_2, . . . , and the distances to the poles P_1, P_2, . . . , and then forming the ratio, so that in the example of Fig. 2.21

$$|H(e^{j\omega T})| = \frac{R_1 R_2}{P_1 P_2 P_3} \tag{2.47}$$

$H(e^{j\omega T})$ also has associated with it a phase, which is given by

$$\varphi = \varphi_1 + \varphi_2 - (\psi_1 + \psi_2 + \psi_3) \tag{2.48}$$

The geometric basis for digital-filter design is thus identical in principle with the geometric basis for continuous-filter design with the following single and important difference: In the continuous filter, frequency is measured along the imaginary axis in the complex s plane whereas, in the digital filter, frequency is measured along the circumference of the unit circle in the z plane.

2.15 DISCRETE NETWORKS WITH NONZERO INITIAL CONDITIONS

The general development of Sec. 2.13 was based on a system initially at rest, that is, all output registers, whose contents were $y(nT)$, $y(nT - T)$,

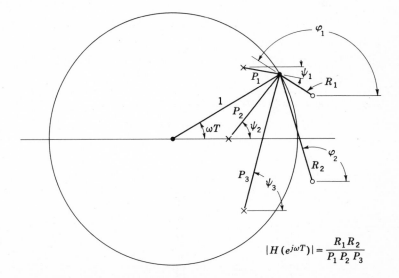

Fig. 2.21 Geometric interpretation of frequency response of digital network.

etc., were set to zero before iteration began.　In order to describe the filter-design techniques of the next chapter, this constraint causes no los in generality, since our filter-design criteria are based on the magnitude function which is, after all, the steady-state response of the filter to sine wave inputs.　However, to deal with the effect of transients, it is clear that initial conditions need to be considered.

As an example, consider (2.32) and attach nonzero values to $y(-T)$ and $y(-2T)$.　Taking the z-transform of both sides of (2.32) gives

$$Y(z) = K_1 \sum_{n=0}^{\infty} y(nT - T)z^{-n} + K_2 \sum_{n=0}^{\infty} y(nT - 2T)z^{-n} + X(z)$$
$$- Lz^{-1}X(z) \quad (2.49)$$

Now, if $m = n - 1$,

$$\sum_{n=0}^{\infty} y(nT - T)z^{-n} = z^{-1} \sum_{m=-1}^{\infty} y(mT)z^{-m} = y(-T) + z^{-1}Y(z)$$

and, if $m = n - 2$,

$$\sum_{n=0}^{\infty} y(nT - 2T)z^{-n} = z^{-2} \sum_{m=2}^{\infty} y(mT)z^{-m}$$
$$= y(-2T) + z^{-1}y(-T) + z^{-2}Y(z)$$

Eq. (2.49) becomes

$$Y(z) = \frac{X(z)(1 - Lz^{-1}) + y(-T)(K_1 + z^{-1}K_2) + K_2 y(-2T)}{1 - K_1 z^{-1} - K_2 z^{-2}}$$
$$(2.50)$$

$Y(z)$ is seen to contain a particular solution due to the input signal $X(z)$ and a transient solution engendered by the initial condition, which eventually decays to zero as long as the poles of $Y(z)$ lie within the unit circle.

2.16　OTHER REALIZATION OF mTH-ORDER SYSTEMS

Equation (2.41) is a ratio of two polynomials in z^{-1} or, if one wishes, in z. It is well known that this rational fraction can be expressed either as a product or as a partial-fraction expansion.　Since poles and zeros of $H(z)$ must either lie on the real axis or occur in complex conjugate pairs, $H(z)$ of (2.41) may be written in either of the two following forms:

$$H(z) = \frac{\displaystyle\prod_{i=1}^{r_1} (z^2 + a_i z + b_i) \prod_{i=1}^{r_2} (z - e_i)}{\displaystyle\prod_{i=1}^{m_1} (z^2 + c_i z + d_i) \prod_{i=1}^{m_2} (z - z_i)} \quad (2.51)$$

or

$$H(z) = \sum_{i=1}^{m_1} \frac{z(A_i z + B_i)}{z^2 + C_i z + D_i} + \sum_{i=1}^{m_2} \frac{z F_i}{z - z_i} \qquad (2.52)$$

Equation (2.51) can be physically interpreted as a *cascade* of second- and first-order systems, and (2.52) can be interpreted as a parallel arrangement of first- and second-order systems. We have deliberately avoided generality; for example, multiple poles would require a more complicated expansion of (2.41). The essential point (and this is completely analogous to the situation for continuous filters) is that any mth-order system can be represented as either a cascade or a parallel (or combination of both) of networks, each of order no higher than 2. Finding these cascade or parallel representations may require finding the roots of mth-order and rth-order polynomials of (2.41). In most instances it proves desirable to realize a given network by means of either cascade or parallel combinations of second-order systems because the latter realizations are less sensitive to the adverse effects associated with finite register length. This important point will be further discussed in Chap. 4.

Another arrangement for the mth-order system is given by the set of equations

$$y_1(nT) = A_{11} y_1(nT - T) + A_{12} y_2(nT - T) + \cdots$$
$$+ A_{1m} y_m(nT - T) + x(nT)$$
$$y_2(nT) = A_{21} y_1(nT - T) + A_{22} y_2(nT - T) + \cdots$$
$$+ A_{2m} y_m(nT - T) \qquad (2.53)$$
$$\cdots \cdots \cdots \cdots \cdots \cdots \cdots \cdots \cdots \cdots \cdots \cdots \cdots$$
$$y_m(nT) = A_{m1} y_1(nT - T) + A_{m2} y_2(nT - T) + \cdots$$
$$+ A_{mm} y_m(nT - T)$$

This set of linear equations can be z-transformed and the poles found by solving the determinant equation

$$\begin{vmatrix} A_{11} - z & A_{12} & \cdots & A_{1m} \\ A_{21} & A_{22} - z & \cdots & A_{2m} \\ \cdots & \cdots & \cdots & \cdots \\ A_{m1} & A_{m2} & \cdots & A_{mm} - z \end{vmatrix} = 0 \qquad (2.54)$$

or equivalently by finding the eigenvalues of the matrix $|A_{ij}|$. Notice that, for an mth-order system, a total of m^2 coefficients are generally needed. A somewhat less general version for which only adjacent sections are cross-coupled, which corresponds to a matrix for which only the main diagonal and the two neighboring diagonals are nonzero, is shown in

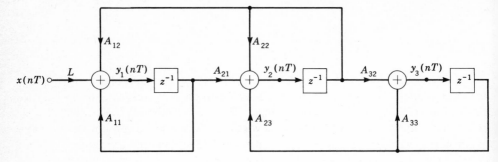

Fig. 2.22   Special case of coupled form.

Fig. 2.22 for $m = 3$.   It is easy to see how Fig. 2.22 can be extended to higher-order systems.

2.17 SUMMARY

The z-transform notation was introduced and applied to the solution of linear difference equations with constant coefficients.   The linear systems described by these equations can also be described in other ways; among them are the frequency-response function, the pole-zero picture in the complex z plane, and the network representation in terms of delays, multipliers, and summations.   It was shown that a second-order difference equation can behave like a resonant circuit and thus has frequency-selective properties.   It was also shown that the z-transform of the output of any linear discrete system can be represented as the product of the z-transforms of the input signal and the network transfer function $H(z)$.   Various digital-network configurations can be derived for the same $H(z)$: the direct form, the canonic form, the parallel form, the cascade form, and the coupled form, such as Fig. 2.22.

PROBLEMS

2.1. Find the solution of Eq. (2.1) when
     (a) $x(nT)$ is a pulse of unity magnitude lasting for M samples.
     (b) $x(nT)$ is zero, but the initial value $y(-T) = 1$.
     (c) $x(nT)$ is a unit step but, after five iterations, the constant multiplier K is changed from 0.8 to 0.95.
2.2. Find the complete solution (transient and steady-state) of the first-order equation (2.1) when the complex exponential $x(nT) = e^{jn\omega T}$ is applied at $nT = 0$.
2.3. Prove that $|H|$ of (2.6) reduces to the magnitude vs. frequency response of the RC filter shown in Fig. 2.5 for small values of T.

2.4. Given $F(z) = \sum\limits_{n=0}^{\infty} f(nT)z^{-n}$

(a) Prove that the z-transform is a linear operation.

(b) Find the z-transform of $f(nT + T)$.

(c) Find the z-transform of $\sum\limits_{k=0}^{n} f(kT)$.

(d) Prove $\lim\limits_{n\to\infty} f(nT) = \lim\limits_{z\to 1} (z - 1)F(z)$.

(e) Prove $f(0) = \lim\limits_{z\to\infty} F(z)$.

(f) Find the z-transform of $nf(nT)$.

2.5. Find the solution to (2.1), using the convolution theorem.

2.6. Using the convolution theorem, find the solution to the difference equation

$$y(nT) = Ky(nT - T) + x(nT) - Lx(nT - T)$$

Assume that $x(nT)$ is a pulse of width M and unity magnitude and that the system was initially at rest.

2.7. Find the inverse z-transform of $\log (z - a)$, $1/(z - a)$, $z/[(z - a)(z - b)]$, where a and b are, in general, complex numbers. What are the constraints on a and b such that the inverse z-transform converges?

REFERENCES

1. James, H. M., N. B. Nichols, and R. S. Phillips: "Theory of Servomechanisms," chap. 5, pp. 231–261, McGraw-Hill Book Company, New York, 1947.
2. Jury, E. I.: "Theory and Application of the z-Transform Method," John Wiley & Sons, Inc., New York, 1964.
3. Kuo, F. F., and J. F. Kaiser (eds.): "Systems Analysis by Digital Computer," John Wiley & Sons, Inc., New York, 1966.
4. Rader, C., and B. Gold: Digital Filter Design Techniques in the Frequency Domain, *Proc. IEEE*, **55**: 149–171 (1967).
5. Ragazzini, J. R., and G. F. Franklin: "Sampled-data Control Systems," McGraw-Hill Book Company, New York, 1958.

3
Frequency-domain Design of Digital Filters

3.1 INTRODUCTION

In Chap. 2 we introduced and developed the z-transform methods and the description of linear difference equations by means of discrete networks. In this chapter we shall apply these methods to the design of certain classes of frequency-selective linear discrete systems, or digital filters, as they are commonly called. These systems have also been called discrete filters and numerical filters. As discussed in Chap. 1, digital filters may be *synthesized* by using direct convolution, linear recursive equations, or fast Fourier transform. On the other hand, the mathematical methods used to *design* digital filters depend strongly on whether the filter impulse response is of finite duration (in which case the filter has no poles, only zeros) or of infinite duration (in which case the filter has, in general, poles and zeros). According to terminology commonly used, recursive filters refer to infinite-duration impulse response and nonrecursive filters refer to finite-duration impulse response. This terminology seems misleading, since the term "recursive" implies that the computation of the output is an explicit function of previous outputs and inputs whereas nonrecursive implies that the output is an explicit function of only previous inputs.

If we accept the above definitions as correct, then, as will be seen in Sec. 3.12, it is possible to synthesize any finite-duration impulse-response filter by using recursive techniques. In Chap. 7, the synthesis of finite-duration impulse-response filters by fast-Fourier-transform techniques and design methods for such filters will be described. In this chapter, except for Sec. 3.12, the design of filters with poles and zeros will be treated. Generally, such filters are synthesized recursively, via linear difference equations. Recently, it has been shown that filters with zeros and poles can also be synthesized by using the fast Fourier transform† but this aspect of the problem will not be treated here.

Prior to the theoretical developments described in this chapter, frequency-selective filters were synthesized by direct convolution. In many practical cases, the impulse response had to be of, say, 100-samples duration to achieve good frequency shaping. This meant that 100 multiplications and additions were required to obtain a single output point. By contrast, to achieve comparable frequency shaping with recursive methods may require only 10 multiplications and additions per output point.

From the above example, it is clear that recursive filtering (wherein an output sample is a function of previous output samples as well as a function of the input) can increase computation speed by an order of magnitude. As we shall see in Chap. 7, nonrecursive-filter designs [for which $h(nT)$ is assumed to be truncated] can be realized by fast-Fourier-transform methods (as opposed to direct computation of the convolution) with great savings in computation time so that, for many problems, nonrecursive design is competitive with recursive design.

Digital filtering is an intrinsic part of many computer simulation programs of dynamical systems. In addition, the decreasing size and cost of digital hardware make it feasible (in certain applications) to design special-purpose hardware for real-time digital filtering. Such filters have several inherent advantages over continuous filters. A greater degree of precision can be attained in the digital realization. A greater variety of digital filters can be built, since certain realization problems (for example, the need for negative inductances) do not arise. No special components are needed to realize digital filters with time-varying coefficients. Aggregates of digital filters should be especially economical in the very-low-frequency band (0.01 to 1 Hz) where the size of passive analog components becomes appreciable.

† B. Gold and K. Jordan, unpublished notes.

3.2 SYNTHESIS OF DIGITAL FILTERS

The design of analog filters can be divided into two parts, the approximation problem and the synthesis problem. The former involves finding a realizable filter transfer function that approximates some chosen ideal. For example, an "ideal" design of a low-pass filter might specify a rectangular magnitude-vs.-frequency-response curve. Since such a function is unrealizable with resistances, capacitances, and inductances, one searches for a suitable approximation. Often this can be done by determining appropriate positions of poles and zeros in the complex frequency domain. Once the poles and zeros are obtained, a synthesis procedure must be found that realizes the transfer function with real components in a practical manner. For example, although it is possible, given the pole and zero pairs, to realize each one with simple resonant circuits and then to connect these circuits in cascade or in parallel, such a procedure involves careful isolation, by means of operational amplifiers, of each pair of complex poles and zeros. For this reason, much theoretical effort has gone into synthesis design procedures for which no isolation is needed.

The approximation problem for digital filters is conceptually no different from that for analog filters. However, since isolation between resonators is inherent between digital elements, the realization can always be done via cascade or parallel combinations of pole and zero pairs, and the elegant procedures of analog-network synthesis do not appear to be needed for digital-filter synthesis. The reader unfamiliar with filter design techniques would surely benefit greatly by studying the approximation problem for analog-filter design; there are many standard treatments of this subject [5, 14, 15].

The above comments should not be interpreted to mean that the specific realization of a given digital filter is unimportant. In Chap. 4 we shall see that the effects of quantization noise are influenced by the configuration chosen. The synthesis problem is real and is affected by many conditions which are, however, different from the conditions imposed in the analog case. After some experience with digital filters, the designer can usually make intelligent decisions as to the synthesis. Throughout this book we are trying to illuminate some of the issues involved, but presently there exists no set of formal synthesis procedures. Some degree of formality has been achieved with respect to the approximation problem, and this will be explicated in the present chapter.

3.3 DISCUSSION OF SEVERAL DIGITAL-FILTER DESIGN TECHNIQUES

Since much information is available on continuous-filter design, a useful approach to digital-filter design involves finding a set of difference equations having a system function $H(z)$ that significantly resembles known analog system functions. The work of Hurewicz [8] provides a technique for doing this in an *impulse-invariant* way. By this we mean that the discrete response to a unit pulse of the derived digital filter will be the samples of the continuous impulse response of the given continuous filter.

Digital filters can be specified via the potential analogy or from a desired squared-magnitude function by using procedures akin to that of the Butterworth and Chebyshev continuous-filter design procedure. This method will be described and realizability conditions discussed in the Appendix at the end of this chapter.

Another technique used by various workers [2, 4, 11, 16] uses conformal mapping to transform a digital-filter design problem into a continuous-filter problem. We shall refer to this technique as the bilinear transformation, although other transformations can be used.

Finally, a technique referred to as *frequency sampling* [12] makes use of the special properties of an elemental digital filter. This elemental filter's frequency-response curve resembles a $(\sin x)/x$ function and has a linear phase-vs.-frequency characteristic. By suitably combining these elemental filters, a simple design technique for a large variety of digital filters can be developed.

Where the same filtering requirements can be adequately met by several different digital filters, the choice among them depends on the speed of execution of the respective computer programs. An important factor in this speed is the number of multiplications (not counting multiplications by such factors as ± 1 or 2^n, which computers can do quickly). Some digital filters are able to meet essentially the same requirements as others even though using fewer multiplications per output sample.

3.4 TECHNIQUE 1: IMPULSE INVARIANCE

We first show that a digital filter with an impulse response equal to the sampled impulse response of a given continuous filter can be derived via the correspondence

$$Y(s) = \sum_{i=1}^{m} \frac{A_i}{s + s_i} \Rightarrow \sum_{i=1}^{m} \frac{A_i}{1 - e^{-s_i T} z^{-1}} = H(z) \tag{3.1}$$

The impulse response $k(t)$ of a continuous filter is defined as the inverse Laplace transform of its system function $Y(s)$, given in general form† by the left side of (3.1). Similarly, the impulse response $h(nT)$ of

† To be completely general, (3.1) should also include terms of the form

$$\frac{A_i}{(s + s_i)^l}$$

which correspond to

$$\left[\frac{(-1)^{l-1}}{(l-1)!} \frac{\partial^{l-1}}{\partial a^{l-1}} \frac{A_i}{1 - e^{-aT} z^{-1}} \right] \Bigg|_{a = s_i}$$

a digital filter is defined as the inverse z-transform of its system function $H(z)$, which can be expressed generally by the right side of (3.1). Thus,

$$k(t) = \mathcal{L}^{-1} \left\{ \sum_{i=1}^{m} \frac{A_i}{s + s_i} \right\} = \sum_{i=1}^{m} A_i e^{-s_i t} \tag{3.2}$$

where $\mathcal{L}^{-1}$ refers to the inverse (one-sided) Laplace transform.

If we desire that $h(nT) = k(t)$, $t = 0, T, 2T, \ldots$, then

$$h(nT) = \sum_{i=1}^{m} A_i e^{-s_i nT} \tag{3.3}$$

Taking the z-transform of (3.3),

$$H(z) = \sum_{n=0}^{\infty} h(nT)z^{-n} = \sum_{i=1}^{m} A_i \sum_{n=0}^{\infty} e^{-s_i nT} z^{-n}$$

$$= \sum_{i=1}^{m} \frac{A_i}{1 - e^{-s_i T} z^{-1}} \tag{3.4}$$

Thus the condition (3.3) that the impulse response of the digital filter be equal to the sampled impulse response of a given continuous filter $Y(s)$ leads to a digital filter defined by (3.4), where all constants A_i and s_i have already been specified from $Y(s)$. By means of the correspondence (3.4), z-transforms can be tabulated [9, 13].

EXAMPLE

The simple one-pole RC low-pass filter is transformed to a digital filter via the correspondence

$$\frac{a}{s + a} \Rightarrow \frac{a}{1 - e^{-aT} z^{-1}} \tag{3.5}$$

System functions of various resonant circuits may be expanded by partial fractions, leading to the correspondences

$$\frac{s + a}{(s + a)^2 + b^2} \Rightarrow \frac{1 - e^{-aT}(\cos bT)z^{-1}}{1 - 2e^{-aT}(\cos bT)z^{-1} + e^{-2aT}z^{-2}}$$

$$\frac{b}{(s + a)^2 + b^2} \Rightarrow \frac{e^{-aT}(\sin bT)z^{-1}}{1 - 2e^{-aT}(\cos bT)z^{-1} + e^{-2aT}z^{-2}} \tag{3.6}$$

3.5 DESIGN OF A DIGITAL LERNER FILTER THAT IS IMPULSE-INVARIANT

Let

$$Y(s) = \sum_{i=1}^{m} \frac{B_i(s + a)}{(s + a)^2 + b_i^2} \tag{3.7}$$

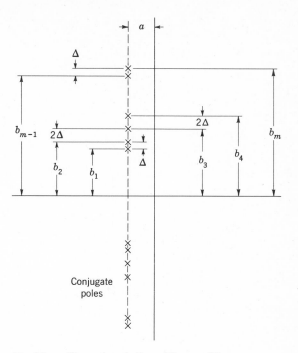

Fig. 3.1 s-Plane description of Lerner filter.

with

$$B_1 = \frac{1}{2} \qquad B_m = \frac{(-1)^{m+1}}{2} \qquad B_i = (-1)^{i+1}$$

$$\text{for } i = 2, \ldots, m-1$$

and the pole positions $-a + b_i$ shown in Fig. 3.1.   It has been shown that Lerner [10] filters have a high degree of phase linearity and reasonably selective passbands.

From the correspondence (3.6), the z-transform of the digital Lerner filter impulse response is

$$H(z) = \sum_{i=1}^{m} \frac{B_i[1 - e^{-aT}(\cos b_i T)z^{-1}]}{1 - 2e^{-aT}(\cos b_i T)z^{-1} + e^{-2aT}z^{-2}} \tag{3.8}$$

The excellent magnitude and phase characteristic of Lerner filters will, however, not be retained in digital realization unless the aliasing effect mentioned in Sec. 2.1 is not significant.   This is the case when the bandwidth of each individual pole is small compared with the sampling frequency.   The invariance of the impulse response may be all that is required in some situations, and this is, of course, not a function of the sampling frequency.

Figure 3.2a and b shows the digital realization of a bandpass Lerner filter with four pole pairs.   Each of the four parallel suboutputs y_i is computed by the difference equation

$$y_i(nT) = e^{-aT} \cos b_i T[2y_i(nT - T) - x(nT - T)]$$
$$- e^{-2aT}y_i(nT - 2T) + x(nT) \qquad i = 1, 2, 3, 4 \quad (3.9)$$

and the output $y(nT)$ is given by†

$$y(nT) = \tfrac{1}{2}y_1(nT) - y_2(nT) + y_3(nT) - \tfrac{1}{2}y_4(nT) \tag{3.10}$$

† The four-pole Lerner filter is especially useful for the design of a filter-bank spectrum analyzer, since two poles can be shared by adjacent filters.   See P. R. Drouilhet and L. M. Goodman, Pole-shared Linear-phase Band-pass Filter Bank, *Proc. IEEE*, **54** (4): (April, 1966).

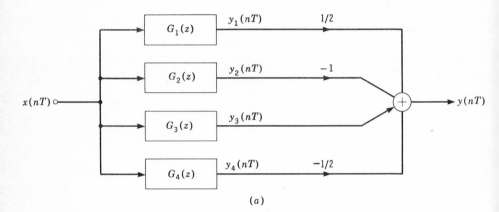

(a)

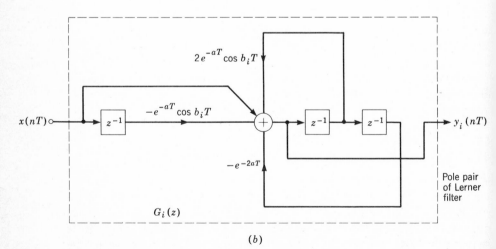

(b)

Fig. 3.2   Four-pole Lerner digital filter.

3.6 GAIN OF DIGITAL RESONATORS

The correspondences (3.6) define two digital resonators which are *impulse invariants* of given continuous resonators.   In practice, digital resonators can be specified without reference to continuous resonators.   Specification consists in placing the pair of complex conjugate poles and, in most cases, a single zero in the z plane, from which the difference equation can be quickly derived.   Since many digital filters are simple cascade or parallel combinations of these resonators, it is important to understand their behavior.

The z-transform of a resonator with poles at $z = re^{\pm j\omega_r T}$ and a zero at q is

$$H(z) = \frac{1 - qz^{-1}}{1 - 2r(\cos \omega_r T)z^{-1} + r^2 z^{-2}} \tag{3.11}$$

The magnitude-vs.-frequency function for (3.11) is $|H(e^{j\omega T})|$ and can be written, by inspection, from Fig. 2.9, using the law of cosines.

$$|H(e^{j\omega T})| = \left\{ \frac{1 + q^2 - 2q \cos \omega T}{[1 + r^2 - 2r \cos (\omega - \omega_r)T][1 + r^2 - 2r \cos (\omega + \omega_r)T]} \right\}^{1/2}$$

$$\tag{3.12}$$

CASE 1 $q = \cos \omega_r T$

For values of r close to unity, the value of $|H(e^{j\omega T})|$ at the resonance ω_r can be approximated by

$$|H(e^{j\omega T})| = \frac{1}{2(1 - r) \sqrt{r}} \tag{3.13}$$

which is independent of ω_r.   Thus this choice of q makes possible the design of an equal-gain bank of resonators (or filters composed of these resonators) covering a wide frequency range.

For narrowband resonators in which r is close to unity, (3.13) shows clearly that the gain at resonance is usually appreciably greater than unity. Knowledge of filter gains is required for determination of appropriate register lengths and to avoid overflow problems.

CASE 2 $q = r \cos \omega_r T$

For this case, the difference equation for (3.11) can be written

$$y(nT) = r \cos \omega_r T[2y(nT - T) - x(nT - T)]$$

$$- r^2 y(nT - 2T) + x(nT) \tag{3.14}$$

Execution of (3.14) requires only two multiplication instructions compared with three multiplications for general q.   This case is thus of special interest for real-time applications and when total computer running time is inordinately great. Sensitivity of the resonant gain with the resonant frequency is greater than for case 1. Thus, case 1 is easier to apply when designing a bank of filters with equal gain at the center frequencies.

CASE 3 $q = 1$

This case yields zero gain when $\omega = 0$, which is often desirable. As in case 2, two multiplications are needed. The gain for $\omega_r T = \pi/2$ is $\sqrt{2}$ times the gain for $\omega_r T$ close to zero, provided that $\omega_r T$ is large compared with $1 - r$.

CASE 4 $q = 0$

In this case, the zero disappears. It is sometimes desirable to design resonators without zeros but with constant gain for $\omega = 0$ independent of ω_r. This is obtained from the digital system function

$$H(z) = \frac{1 - 2r \cos \omega_r T + r^2}{1 - 2r \cos \omega_r T z^{-1} + r^2 z^{-2}} \tag{3.15}$$

3.7 DESIGN OF DIGITAL FILTERS FROM CONTINUOUS FILTERS THAT HAVE ZEROS AT INFINITY

A large class of analog filters are defined by system functions of the form

$$Y(s) = \frac{1}{\displaystyle\prod_{i=1}^{m} (s + s_i)} \tag{3.16}$$

where the denominator is a product. Such filters have m poles at finite values of s and an mth-order zero for infinite s. Included in this category are Butterworth, Chebyshev, and Bessel filters.

In order to design impulse-invariant digital filters based on (3.16), the procedure outlined in Sec. 3.3 can be used: $Y(s)$ is expanded in partial fractions, the A_i are found, and $H(z)$ is obtained from the correspondence (3.1). In general, this causes zeros to appear in $H(z)$, although there were no finite zeros in $Y(s)$. However, when the poles at $-s_i$ are close to the imaginary axis in the s plane, so that $e^{-s_i T}$ is close to unity, the zeros of $H(z)$ can be ignored, and $H(z)$ may be approximated by

$$H(z) = \frac{1}{\displaystyle\prod_{i=1}^{m} (1 - e^{-s_i T} z^{-1})} \tag{3.17}$$

In practice, digital bandpass filters several hundred hertz wide, of the Bessel, Butterworth, or Chebyshev type, have been successfully programmed for 10,000-Hz sampling rates by using the form (3.17).

EXAMPLE: THREE–POLE BUTTERWORTH LOW–PASS FILTER

The system function of the continuous filter is

$$Y(s) = \frac{s_1 s_2 s_3}{(s + s_1)(s + s_2)(s + s_3)} \tag{3.18}$$

with

$$s_1 = \omega_c \qquad s_2 = \tfrac{1}{2}(1 + j\sqrt{3})\omega_c \qquad s_3 = \tfrac{1}{2}(1 - j\sqrt{3})\omega_c$$

where ω_c is the cutoff frequency, defined by $|Y(j\omega_c)| = 0.707$. Expansion of (3.18) into partial fractions and the correspondence (3.1) lead to the z-transform

$$H(z) = \omega_c \left[\frac{1}{1 - e^{-\omega_c T}z^{-1}} + \frac{-1 + \eta z^{-1}}{1 - \beta z^{-1} + e^{-\omega_c T}z^{-2}} \right] \tag{3.19}$$

with η and β defined, along with the diagrammatic representation of (3.19), in Fig. 3.3.

If a cascade representation is desired, (3.19) can be written to give

$$H(z) = \frac{Cz^{-1} + Dz^{-2}}{(1 - e^{-\omega_c T}z^{-1})\{1 - 2e^{-\omega_c T/2}[\cos(\sqrt{3}/2)\omega_c T]z^{-1} + e^{-\omega_c T}z^{-2}\}} \tag{3.20}$$

with

$$C = \omega_c \left[e^{-\omega_c T} + e^{-\omega_c T/2} \left(\frac{1}{\sqrt{3}} \sin\frac{\sqrt{3}}{2}\omega_c T - \cos\frac{\sqrt{3}}{2}\omega_c T \right) \right]$$

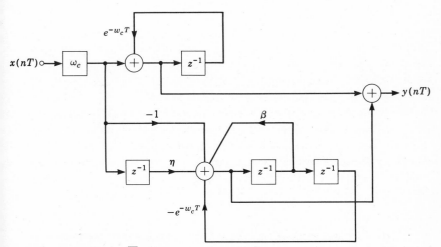

$\beta = 2 \exp[-\omega_c T/2] \cos(\sqrt{3}\,\omega_c T/2)$

$\eta = \exp[-\omega_c T/2] [\cos(\sqrt{3}\,\omega_c T/2) + (\sin(\sqrt{3}\,\omega_c T/2))/\sqrt{3}]$

Fig. 3.3 Three-pole Butterworth low-pass digital filter.

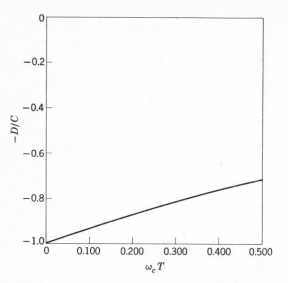

Fig. 3.4 Variation of position of zero of impulse-invariant three-pole Butterworth low-pass digital filter.

and

$$D = \omega_c \left[e^{-\omega_c T} - e^{-3\omega_c T/2} \left(\frac{1}{\sqrt{3}} \sin \frac{\sqrt{3}}{2} \omega_c T + \cos \frac{\sqrt{3}}{2} \omega_c T \right) \right]$$

$H(z)$ is seen to have two zeros, one at $z = 0$ and the other at $z = -D/C$. This zero is on the real axis and increases as $\omega_c T$ is increased from zero. Note that the denominator of (3.20) can be written by inspection of the poles of the continuous filter, since an s-plane pole transforms directly into a z-plane pole via $z = e^{sT}$.

For small values of $\omega_c T$, the zero of $H(z)$ is approximately at -1 and thus has but a small effect on the bandpass characteristics of the three-pole Butterworth filter. Figure 3.4 shows a plot of $-D/C$ versus $\omega_c T$. We see, for example, that the zero moves to -0.94 for $\omega_c T = 0.1$, which still introduces but slight distortion in the passband of the cascade approximation.

3.8 REVIEW OF BUTTERWORTH, CHEBYSHEV, AND ELLIPTIC FILTERS

In this section, we review briefly design procedures for continuous Butterworth, Chebyshev, and elliptic filters, which will be needed in the later discussion of digital-filter design. More complete treatments are available in standard texts [5, 14, 15].

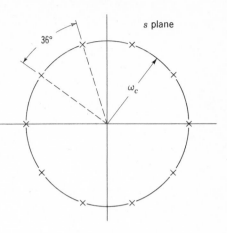

Fig. 3.5 Poles of Butterworth filter in s plane.

The Butterworth filter can be specified by the relationship

$$|F(j\omega)|^2 = \frac{1}{1 + (\omega/\omega_c)^{2n}} \tag{3.21}$$

where ω_c is a cutoff frequency and $|F(j\omega)|^2$ is the squared magnitude of a filter transfer function. The poles of (3.21) lie equally spaced on a circle of radius ω_c in the s plane, as shown in Fig. 3.5. For n odd, there are poles at angles of 0 and π; for n even, the first pole occurs at an angle of $\pi/2n$. It can be shown that the desired transfer function $F(s)$ is a rational function with constant numerator and denominator determined by the left-half poles of Fig. 3.5. Plots of $|F(j\omega)|$ of (3.21) for several values of n are shown in Fig. 3.6. From these plots the selectivity properties of the Butterworth filter become clear.

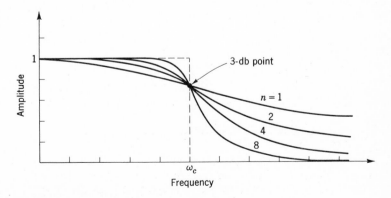

Fig. 3.6 Butterworth frequency responses.

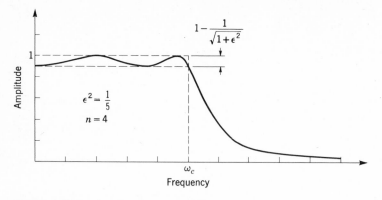

Fig. 3.7 Chebyshev frequency response.

The Chebyshev filter is specified by

$$|F(j\omega)|^2 = \frac{1}{1 + \epsilon^2 V_n^2(\omega/\omega_c)} \tag{3.22}$$

where $V_n(x)$ is a Chebyshev polynomial of order n which can be generated by the recursion formula

$$V_{n+1}(x) - 2xV_n(x) + V_{n-1}(x) = 0 \tag{3.23}$$

with

$$V_1(x) = x \quad \text{and} \quad V_2(x) = 2x^2 - 1$$

The Chebyshev polynomial has the property of equal ripple over a given range of x which, with added specification of ϵ, leads to a magnitude function of a form given by Fig. 3.7, an equal ripple in the passband, and a monotonic decay in the stopband. The ripple amplitude δ is given by

$$\delta = 1 - \frac{1}{\sqrt{1 + \epsilon^2}}$$

The poles of (3.22) lie on an ellipse which can be determined totally by specifying ϵ, n, and ω_c. Figure 3.8 shows this ellipse, the vertical and horizontal apices being given by $b\omega_c$, $a\omega_c$, where

$$b, a = \tfrac{1}{2}[(\sqrt{\epsilon^{-2} + 1} + \epsilon^{-1})^{1/n} \pm (\sqrt{\epsilon^{-2} + 1} + \epsilon^{-1})^{-1/n}]$$

where the b is given for the plus sign and a for the minus sign. The poles on the ellipse may be geometrically related to the poles of two Butterworth circles of radii $a\omega_c$ and $b\omega_c$. The vertical position of the ellipse pole is equal to the vertical position of the pole on the large circle, and the hori-

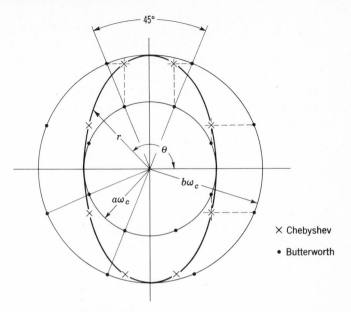

Fig. 3.8 Poles of Chebyshev filter in s plane.

zontal position is that of the horizontal position of the small-circle pole.

The elliptic filter is based on the properties of the Jacobian elliptic function, first analyzed by Jacobi [7], which has since been treated extensively in several mathematical treatises [1, 17]. For our purpose, it is

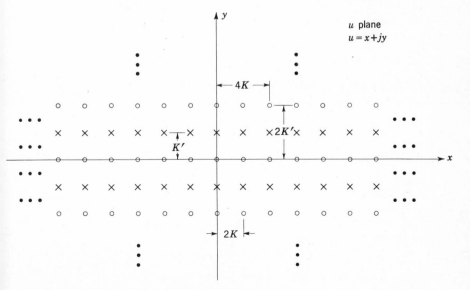

Fig. 3.9 Poles and zeros of Jacobian elliptic function.

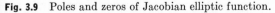

important to know that the Jacobian elliptic function, denoted by sn u, is a doubly periodic function of the complex variable u, analytic in the u plane except at the simple poles of the function. This is in evidence in Fig. 3.9; since sn u is doubly periodic, the basic pattern of two zeros and two poles is repeated infinitely along both x and y axes. The residues associated with each pole pair (adjacent poles along any horizontal path) are equal and opposite.

The function sn u is sometimes written sn (u,k) to indicate the presence of the parameter k. Analytically, if $y = $ sn (u,k), then

$$u = \int_0^y \frac{dt}{(1 - t^2)^{1/2}(1 - k^2 t^2)^{1/2}} = \text{sn}^{-1} y \qquad (3.24)$$

that is, y is the inversion of an incomplete elliptic integral of the first kind. The parameters $K(k)$ and $K'(k)$ shown in Fig. 3.9 are given by

$$K(k) = \text{sn}^{-1} 1 \qquad K'(k) = K(k') \qquad \text{where } k' = (1 - k^2)^{1/2} \qquad (3.25)$$

The doubly periodic nature of sn u and its pole-zero pattern make it possible to find a squared-magnitude function that has the desirable properties of equal ripple in both passband and stopband, as shown in Fig. 3.10. To demonstrate this, we consider the function

$$T^2(u) = \frac{1}{1 + \epsilon^2 \text{ sn}^2 (u,k_1)} \qquad (3.26)$$

where $k_1 = \epsilon/\sqrt{A^2 - 1}$ and ϵ and A are the same parameters as shown in Fig. 3.10. The pole-zero pattern of $T^2(u)$ is shown in Fig. 3.11a. The poles of sn u become double zeros of $T^2(u)$. The poles of $T^2(u)$ are found by solving the equation

$$1 + \epsilon^2 \text{ sn}^2 u = 0$$

and this yields the array of poles shown in Fig. 3.11a. The pattern of Fig. 3.11a is also repeated doubly periodically. The following analysis,

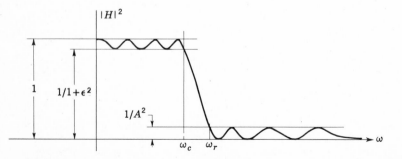

Fig. 3.10 Elliptic-filter frequency response.

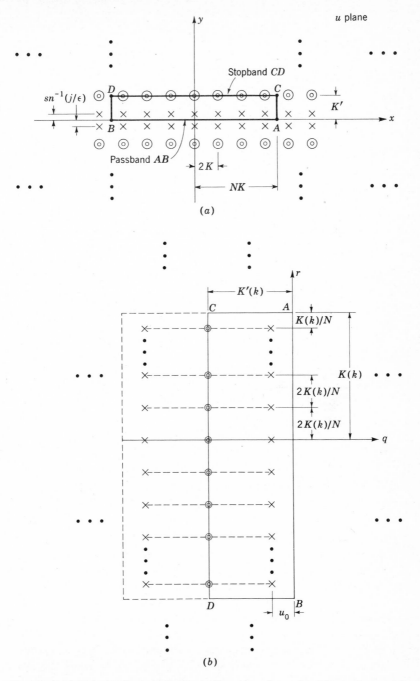

Fig. 3.11 (a) Poles and zeros of elliptic filter in u plane; (b) poles and zeros of elliptic filter in λ plane for N odd.

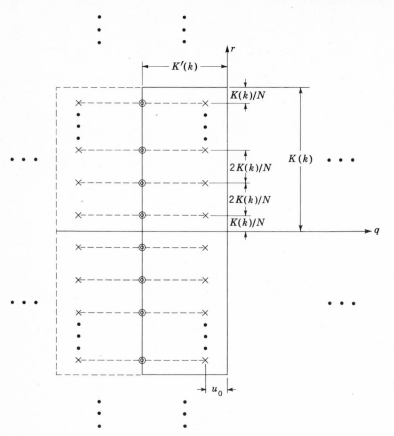

Fig. 3.12 Poles and zeros of elliptic filter in λ plane for N even.

for odd values of N, is based on Fig. 3.11b. For even values of N, Fig. 3.12 holds, and the analysis is similar and will not be explicitly reviewed.

By examining the points $u = K$ and $u = K + jK'$, we can explicitly prove the ripple relationships of Fig. 3.10. From symmetry considerations of Fig. 3.11a, we can define a passband and a stopband. The passband is on the x axis, and the stopband is the horizontal line through the zeros above the x axis. Around the rectangle shown in Fig. 3.11a, we make the transition from passband to stopband along the vertical lines AC and, for negative frequencies, BD. Furthermore, the point $u = K(k_1)$ corresponds to a minimum value of $T^2(u)$ in the passband, and the point $u = K(k_1) + jK'(k_1)$ corresponds to a maximum in the stopband. Using the relations sn $K = 1$ and sn $(K + jK') = 1/k_1$ leads directly to the ripple characteristics of Fig. 3.10. However, this figure gives a squared-magnitude function along a line going to infinity, whereas Fig. 3.11a gives the function along a rectangle in the u plane. To complete the picture, the rectangle must be mapped onto the $j\omega$ axis in the s plane. This mapping is conveniently done in two steps. First, we make the transformation

$$\lambda = j\,\frac{K(k)u}{NK(k_1)} \qquad \text{with } k = \frac{\omega_c}{\omega_r}$$

which converts the pattern of Fig. 3.11a into that of Fig. 3.11b. Finally, the rectangle of Fig. 3.11b must be mapped onto the s plane as the $j\omega$ axis. The reader can verify that if we let

$$N = \frac{K'(k_1)K(k)}{K(k_1)K'(k)} \tag{3.27}$$

and make the conformal transformation

$$s = j\omega_c \,\text{sn}\,(-j\lambda,k) \tag{3.28}$$

we obtain the mapping shown in the following table:

λ plane	s plane
0	0
$\pm jNK(k)$	$\pm j\omega_c$
$\pm jNK(k) - K'(k)$	$\pm j\omega_r$
$- K'(k)$	$\pm j\,\infty$

Furthermore, it can be verified that the rectangle $ABCD$ of Fig. 3.11b does indeed map onto the $j\omega$ axis in the s plane. The precise positions of the poles and zeros in the s plane are obtained from (3.28), which may be expressed in terms of two real transformations; if $s = \sigma + j\omega$ and $\lambda = q + jr$, then

$$\sigma = \omega_c \,\frac{\text{sn}\,(q,k')\,\text{cn}\,(q,k')\,\text{cn}\,(r,k)\,\text{dn}\,(r,k)}{1 - \text{sn}^2\,(q,k')\,\text{dn}^2\,(r,k)}$$

$$\omega = \omega_c \,\frac{\text{sn}\,(r,k)\,\text{dn}\,(q,k')}{1 - \text{sn}^2\,(q,k')\,\text{dn}^2\,(r,k)} \tag{3.29}$$

The cn and dn functions are easily obtained from the formulas

$$\text{cn}^2\,u = 1 - \text{sn}^2\,u \qquad \text{dn}^2\,u = 1 - k^2\,\text{sn}^2\,u$$

and u_0 in Fig. 3.11b can be written

$$u_0 = -j\,\frac{K(k)}{NK(k_1)}\,\text{sn}^{-1}\frac{j}{\epsilon} \tag{3.30}$$

Let us comment on the differences between Butterworth, Chebyshev, and elliptic filters. All these filters are approximations to a desired rectangular passband. The Butterworth filter achieves this via a monotonic amplitude-vs.-frequency characteristic. By allowing a ripple in the passband, the Chebyshev filter, using the same number of poles and zeros, can achieve sharper cutoff. Elliptic filters yield even sharper cutoff than Chebyshev for the same network complexity but allow both passband and stopband ripple. The choice of filter thus depends on the application, the size considerations, and the time one is willing to spend designing the filter.

Before delving further into the details of digital-filter design techniques, let us digress briefly for a few general remarks. To a great extent, the theory already presented and to be presented in this chapter appears to be an adaptation of analog-filter theory to the digital-filter problem. Perhaps a somewhat broader viewpoint is that filter theory is of a general nature and can be applied to specific situations. For example, the discussion leading to (3.26) makes no mention of specific filters; (3.26) simply specifies a function with desirable properties on a rectangle in the u plane of Fig. 3.11a. To convert this function into one that is useful for an analog-filter synthesis requires a transformation that maps this rectangle onto the $j\omega$ axis in the s plane. If one wanted, instead, to use this same function for digital-filter synthesis, it would be necessary to map the rectangle of Fig. 3.11a onto the unit circle in the z plane. From this point of view, one can apply "general" filter theory directly to either analog or digital realizations. Similarly, as we shall see in Sec. 3.9, squared-magnitude functions having a form suitable for direct application to digital-filter design can be specified. On the other hand, the reader who is well versed in continuous-filter design may find it convenient to use the techniques of Sec. 3.10 to go directly from analog to digital design.

3.9 TECHNIQUE 2: DIGITAL-FILTER SPECIFICATION FROM SQUARED-MAGNITUDE FUNCTION

We have seen in Sec. 3.8 that the Butterworth and Chebyshev filters are specified by choosing suitably selective squared-magnitude functions such as in (3.21) and (3.22) or (3.26). The same procedure is possible for digital filters and is described in this section.

It has now been established that the digital-filter system function $H(z)$ is a rational function in z^{-1}; it follows that $H(z)$ for z on the unit circle is a rational function of $e^{j\omega T}$. Thus, the squared magnitude $|H(e^{j\omega T})|^2$ can always be expressed as the ratio of two trigonometric functions of ωT.

An example of a squared-magnitude function suitable for low-pass

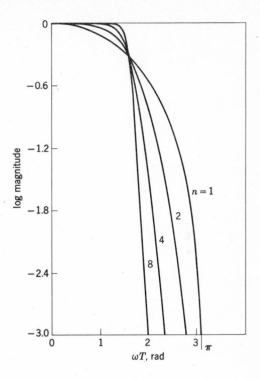

Fig. 3.13 Frequency responses of digital Butterworth filters.

filtering is

$$|H(e^{j\omega T})|^2 = \cfrac{1}{1 + \cfrac{\tan^{2n}(\omega T/2)}{\tan^{2n}(\omega_c T/2)}} \tag{3.31}$$

Equation (3.31) is plotted in Fig. 3.13 for $\omega_c T = \pi/2$ for several values of n. The curves obtained are similar to those of the Butterworth filter of (3.21) plotted in Fig. 3.6. The cutoff frequency ω_c plays the same role in both the continuous and digital cases.

Letting $z = e^{j\omega T}$, (3.31) may be rewritten

$$|H(z)|^2 = \frac{\tan^{2n}(\omega_c T/2)}{\tan^{2n}(\omega_c T/2) + (-1)^n[(z-1)/(z+1)]^{2n}} \tag{3.32}$$

We see that (3.32) is a rational function in z which has a zero of order $2n$ at $z = -1$. The poles are found by substituting, in (3.32),

$$p = \frac{z-1}{z+1} \tag{3.33}$$

from which we can ascertain that the $2n$ poles of $|H(p)|^2$ are uniformly spaced around a circle of radius $\tan(\omega_c T/2)$ in the p plane. The poles in

z are then readily found by the transformation inverse to (3.33), namely,

$$z = \frac{1 + p}{1 - p} \tag{3.34}$$

Letting $p = x + jy$ and $z = u + jv$, we find from (3.34) the component relations

$$u(x,y) = \frac{1 - x^2 - y^2}{(1 - x)^2 + y^2} \qquad v(x,y) = \frac{2y}{(1 - x)^2 + y^2} \tag{3.35}$$

The circle containing the poles in the p plane satisfies the equation

$$x^2 + y^2 = \tan^2 \frac{\omega_c T}{2} \tag{3.36}$$

From (3.35) and (3.36) it can be shown that the circle maps into a circle in the z plane centered at (u_c, v_c) with radius ρ:

$$
\begin{aligned}
u_c &= \frac{1 + \tan^2 (\omega_c T/2)}{1 - \tan^2 (\omega_c T/2)} = \sec \omega_c T \qquad v_c = 0 \\
\rho &= \frac{2 \tan (\omega_c T/2)}{1 - \tan^2 (\omega_c T/2)} = \tan \omega_c T
\end{aligned}
\tag{3.37}
$$

For odd values of n, the $2n$ poles in the p plane have the x and y coordinates

$$
\begin{aligned}
x_m &= \tan \frac{\omega_c T}{2} \cos \frac{m\pi}{n} \\
y_m &= \tan \frac{\omega_c T}{2} \sin \frac{m\pi}{n}
\end{aligned}
\qquad m = 0, 1, \ldots, 2n - 1 \tag{3.38}
$$

For even values of n, the coordinates are

$$
\begin{aligned}
x_m &= \tan \frac{\omega_c T}{2} \cos \frac{2m + 1}{2n} \pi \\
y_m &= \tan \frac{\omega_c T}{2} \sin \frac{2m + 1}{2n} \pi
\end{aligned}
\qquad m = 0, 1, \ldots, 2n - 1 \tag{3.39}
$$

From (3.38) and (3.39) the corresponding poles in the z plane are computed to be

$$
\begin{aligned}
u_m &= \frac{2[1 - \tan^2 (\omega_c T/2)]}{1 - 2 \tan (\omega_c T/2) \cos (m\pi/n) + \tan^2 (\omega_c T/2)} \\
v_m &= \frac{2 \tan (\omega_c T/2) \sin (m\pi/n)}{1 - 2 \tan (\omega_c T/2) \cos (m\pi/n) + \tan^2 (\omega_c T/2)}
\end{aligned}
$$
$$m = 0, 1, \ldots, 2n - 1 \tag{3.40}$$

Replacing $m\pi/n$ by $(2m + 1)\pi/2n$ yields equivalent formulas for n even.

EXAMPLE

Find the poles and zeros of the squared-magnitude function of a low-pass filter with 3-db attenuation at 1,250 Hz and with at least 20-db attenuation at 2,000 Hz.   Let the sampling rate be 10,000 Hz.

The cutoff frequency of 1,250 Hz corresponds to $\omega_c T = 45°$.   The frequency 2,000 Hz corresponds to $\omega T = 72°$.

The squared-magnitude function (3.31) becomes

$$|H(e^{j\omega T})|^2 = \frac{1}{1 + \dfrac{\tan^{2n}(\omega T/2)}{\tan^{2n}(\pi/8)}} \tag{3.41}$$

The appropriate value of n in (3.41) is $n = 4$, obtained by setting $\omega T = 72°$ and $|H(e^{j\omega T})|^2$ to 0.01, thus satisfying the 20-db attenuation condition.   The eight poles in the p plane are found from (3.39).   Equation (3.40) can now be used to find the z-plane poles shown in Fig. 3.14; this figure also shows the $2n$ zeros located at $z = -1$, which are directly derivable from (3.32).   The squared-magnitude function is thus completely specified as pole-zero placements in the z plane.

In the Appendix of this chapter, an analysis will be made of the necessary relations between an assumed squared-magnitude function and the digital filter specified by that function.   It is shown that, in order for a squared-magnitude function to be realizable, any pole inside the unit circle (for example, z_4 in Fig. 3.14) must have a mate of inverse magnitude

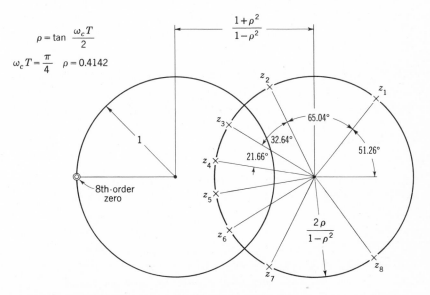

Fig. 3.14   Poles and zeros of digital Butterworth filters.

and the same angle outside the unit circle. Thus, if $z_4 = re^{j\psi}$, there must be a pole (in the case, z_8) given by $(1/r)e^{j\psi}$. In addition, all poles must occur in complex conjugate pairs. Therefore, the digital filter derived from Fig. 3.14 has the conjugate poles z_4, z_5 and z_3, z_6.

The above argument holds for the zeros as well, including the special case of zeros on the real axis. In Fig. 3.14, all eight zeros occur at $z = -1$. The derived filter has four zeros at $z = -1$.

If the squared-magnitude function is given by

$$|H(e^{j\omega T})|^2 = \frac{1}{1 + \epsilon^2 V_n{}^2 \left(\dfrac{\tan(\omega T/2)}{\tan(\omega_c T/2)} \right)} \tag{3.42}$$

it can be shown that the poles of $p = (z - 1)/(z + 1)$ lie on an ellipse in the p plane that has the same properties as the Chebyshev ellipse of Fig. 3.8. Using the notation of Sec. 3.8 and Fig. 3.14, the p-plane components can be written

$$x = a \tan \frac{\omega_c T}{2} \cos \theta$$
$$\tag{3.43}$$
$$y = b \tan \frac{\omega_c T}{2} \sin \theta$$

Substituting (3.43) into (3.35) yields

$$u = \frac{2[1 - a \tan(\omega_c T/2) \cos \theta]}{[1 - a \tan(\omega_c T/2) \cos \theta]^2 + b^2 \tan^2(\omega_c T/2) \sin^2 \theta} - 1$$
$$\tag{3.44}$$
$$v = \frac{2b \tan(\omega_c T/2) \sin \theta}{[1 - a \tan(\omega_c T/2) \cos \theta]^2 + b^2 \tan^2(\omega_c T/2) \sin^2 \theta}$$

Figure 3.15 shows the z-plane mapping for $a \tan(\omega_c T/2) = 0.5$ and $b \tan(\omega_c T/2) = 1$. The ellipse of Fig. 3.8 maps into the cardioid-like curve of Fig. 3.15, and the inner circle of Fig. 3.8 maps into the right-hand circle of Fig. 3.15. The outer circle of Fig. 3.8 maps into a circle of infinite radius, shown by the straight line of Fig. 3.15. The points shown on the mapped ellipse of Fig. 3.15 are computed from (3.44).

3.10 TECHNIQUE 3: DESIGN OF DIGITAL FILTERS USING BILINEAR TRANSFORMATION OF CONTINUOUS-FILTER FUNCTION

In the previous section it was shown how the poles and zeros of a digital filter with a suitable squared-magnitude function can be found. This was completely analogous to the mathematics of analog-filter design, which has reached a considerable degree of sophistication. For many digital-filter design problems, the substitution of (3.33) transforms the digital-filter design problem into a problem that can be recognized as identical to

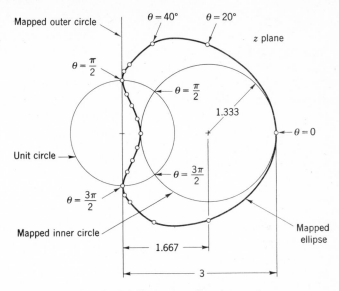

Fig. 3.15 Mapping of Chebyshev filter into z plane.

an already solved analog-filter design problem. For example, (3.33) transformed the problem of finding the pole locations for (3.32) and (3.42) into the already-solved problems of finding the pole locations of analog Butterworth and Chebyshev filters. This observation leads to another digital-filter design approach, using design directly in the s plane.

Suppose we have a stable analog filter described by $H(s)$. Its frequency response is found by evaluating $H(s)$ at points on the imaginary axis of the s plane. If, in the function $H(s)$, s is replaced by a rational function of z which maps the imaginary axis of the s plane onto the unit circle of the z plane, then the resulting $H'(z)$, evaluated along the unit circle, will take on the same set of values $H(s)$ evaluated along the imaginary axis.

This does not mean the functions are the same, for the frequency scales are distorted relative to one another. This is illustrated by the simplest rational function which maps the $j\omega$ axis onto the unit circle

$$s \rightarrow \frac{z - 1}{z + 1} \tag{3.45}$$

Let the analog frequency variable be ω_A, and let the digital frequency variable be $\omega_D T$. Then the functions $H(\omega_A)$, $H'(\omega_D T)$ take on the same values for

$$\omega_A = \tan \frac{\omega_D T}{2} \tag{3.46}$$

Note that the transformation (3.45) leads to a ratio of polynomials in z. Since it maps the left half of the s plane onto the inside of the unit circle, we can be sure that it will always yield for $H'(z)$ a realizable, stable digital filter.

Equations (3.45) and (3.46) yield a technique for designing a digital filter by analog techniques. The procedure is as follows:

1. Note the critical frequencies and ranges (passband or stopband, maximum attenuation point, etc.) of the desired digital filter, and call them $\omega_{D_i}T$. Compute a new set of frequencies, ω_{A_i}, by

$$\omega_{A_i} = \tan \frac{\omega_{D_i}T}{2} \tag{3.47}$$

2. Design a transfer function $H(s)$ with the properties of the digital filter at the new frequencies and ranges. There is no need to synthesize $H(s)$.

3. Replace s by $(z - 1)/(z + 1)$ in $H(s)$, and perform the algebra necessary to express the resulting $H'(z)$ as a ratio of polynomials; this yields the desired digital filter.

This technique is illustrated by the following example, in which we design a digital filter for a 10-kHz sampling rate, which is flat to 3 db in the passband of 0 to 1,000 Hz and which is more than 10 db down at frequencies beyond 2,000 Hz. The filter must be monotonic in passband and stopband.

From Fig. 3.6 we see that a Butterworth filter can meet the above requirement in the analog domain. The critical frequencies are

$$\omega_{D_1}T = 2\pi \times 0.1$$

and

$$\omega_{D_2}T = 2\pi \times 0.2$$

1. Compute ω_{A_1}, ω_{A_2}:

$$\omega_{A_1} = \tan \frac{2\pi \times 0.1}{2} = 0.3249$$

$$\omega_{A_2} = \tan \frac{2\pi \times 0.2}{2} = 0.7265$$

2. Design a Butterworth filter with 3-db point at $\omega_c = \omega_{A_1} = 0.3249$, $\omega_{A_2}/\omega_c = 2.235$. To find the order n, solve $1 + (2.235)^{2n} = 10$ and obtain $n = 2$. A second-order Butterworth filter with $\omega_c = 0.3249$

has poles at $s = 0.3249 \times (-0.707 \pm j0.707) = -0.23 \pm j0.23$, and no zeros.

$$H(s) = \frac{s_1 s_2}{(s + s_1)(s + s_2)} = \frac{2 \times (0.23)^2}{(s + 0.23)^2 + (0.23)^2}$$

$$= \frac{0.1058}{s^2 + 0.46s + 0.1058}$$

3. Replace s by $(z - 1)/(z + 1)$, yielding $H'(z)$.

$$H'(z) = \frac{0.1058}{(z - 1)^2/(z + 1)^2 + 0.46(z - 1)/(z + 1) + 0.1058}$$

$$H'(z) = \frac{0.1058(z^2 + 2z + 1)}{1.5658z^2 - 1.7884z + 0.6458} = \frac{0.067569(z^2 + 2z + 1)}{z^2 - 1.14216z + 0.412441}$$

This is the required digital filter. It requires three multiplications per output point or, if dc gain can be tolerated, only two multiplications.

EXAMPLE

Design a digital filter passing from 0 to 100 Hz with $\frac{1}{2}$-db ripple, which falls off monotonically to at least -19 db at 183 Hz. Use a 1,000-Hz sampling rate.

1. The critical frequencies 100 and 183 Hz are transformed to analog frequencies:

$$\omega_c = \tan \frac{2\pi \times 100}{2 \times 1,000} = \tan 18° = 0.32492$$

$$\omega_s = \tan \frac{2\pi \times 183}{2 \times 1,000} = 0.6498 \approx 2\omega_c$$

2. We now design an analog filter of the Chebyshev variety. One-half-decibel ripple corresponds to $\epsilon^2 = 0.1220184$. To find the required order we solve $1 + \epsilon^2 V_n{}^2(\omega_a/\omega_c) \geq 10^{1.9}$. The lowest order n satisfying this relationship is $n = 3$. A unity-bandwidth, $\frac{1}{2}$-db-ripple Chebyshev filter is

$$H_1(s) = \frac{\text{constant}}{s^3 + 1.252913s^2 + 1.5348954s + 0.7156938}$$

Replacing s by s/ω_c yields

$$H(s) = \frac{0.0255842155}{s^3 + 0.4127346s^2 + 0.16656307s + 0.0255842155}$$

where the constant has been adjusted for unity gain at $s = 0$.

3. We replace s by $(z - 1)/(z + 1)$, giving for the digital filter desired,

after multiplying numerator and denominator by $(z + 1)^3$,

$$H'(z) = \frac{0.0159414914(z^3 + 3z^2 + 3z + 1)}{z^3 - 1.974860236z^2 + 1.524277838z - 0.45376786}$$

which is the required digital-filter design.

It is worth noting a useful geometric interpretation of step 3. Replacing s by $(z - 1)/(z + 1)$ is a mapping of points in the s plane onto points in the z plane. The accompanying short table gives the correspondence of some critical points in the s and z planes. A very similar

s plane	z plane
$0 + j0$	$1 + j0$
∞	$-1 + j0$
$0 + j1$	$0 + j1$
$0 - j1$	$0 - j1$
$-1 + j0$	$0 + j0$
Point on real axis	Point on real axis
Point on imaginary axis	Point on unit circle
Point on any line	Point on circle passing through $-1 + j0$

mapping has been found useful in some applications, and graph paper that performs the transformation can be purchased under the name Smith chart. To perform the mapping of our application, we take a conventional Smith chart and rotate it 180°, giving a chart like Fig. 3.16.

The location of any point $-a + jb$ in the s plane (left half) is used to find the corresponding location in the z plane as follows:

1. Locate a, a point along the centerline. All points along the circle passing through this point correspond to points in the s plane with real part equal to $-a$.
2. Locate b, a point along the perimeter of the outer circle. For $b > 0$, use the top semicircle; for $b < 0$, use the lower semicircle. The circular arc passing through this point corresponds to points in the s plane with imaginary part b.
3. The intersection of the circle with the circular arc is the point in the z plane corresponding to $-a + jb$ in the s plane.

The Smith chart is useful when the function in the s plane is known in terms of poles and zeros or poles and residues, especially when the z-plane representation is desired in that form.

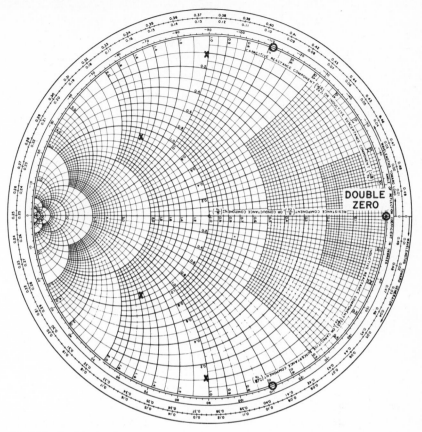

Fig. 3.16 Smith chart.

EXAMPLE

Design a digital high-pass filter with stopband (0 to 500 Hz) attenuation greater than 36 db and passband (above 660 Hz) ripple of 1.25 db. Sampling rate is 2.5 kHz.

1. The critical frequencies are transformed to analog, giving, as the stopband limit, 0.72654 rad/sec and, as the passband limit, 1.09 rad/sec.
2. The specifications are met with a fourth-order elliptic filter. Only the result is given here. The poles are at $s = -1.1915812 \pm j1.5528835$ and $-0.11078815 \pm j1.09445605$, and the zeros are at $s = 0$ (double) and $\pm j0.69117051$. These are located on a Smith chart in Fig. 3.16. By using a ruler and protractor, the zeros in the z plane are found to be at $z = 1$ (double) and $e^{\pm j69.3°}$, and the poles are found to be at $z = 0.586e^{\pm j131.6°}$ and $0.895e^{\pm j90.6°}$. The function

of z defining the filter is thus

$$H(z) = \frac{(z - 1)^2(z^2 - 0.707z + 1)}{(z^2 + 0.777z + 0.3434)(z^2 + 0.01877z + 0.801)}$$

A prime difficulty with the Smith-chart method, illustrated by this example, is that the poles cannot be located with great accuracy. In compensation, a good deal of insight into the operation of the digital filter is gained in the course of the design. Where greater accuracy is needed, the z-plane poles can be computed from the s-plane poles by using $z = (1 + s)/(1 - s)$.

The transformation $s \to (z - 1)/(z + 1)$ is not the only rational function of z which maps the imaginary axis of the s plane onto the unit circle. For example, another such transformation is

$$s \to \frac{z^2 - 2z \cos \psi_0 T + 1}{z^2 - 1} \tag{3.48}$$

For this transformation the imaginary axis of the s plane maps onto the top arc of the unit circle and also onto the bottom arc. The origin of the s plane maps onto the two points $e^{\pm j\psi_0 T}$.

Thus (3.48) maps a frequency function $H(\omega_A)$ into $H'(\omega_D T)$ with

$$\omega_A = \frac{\cos \psi_0 T - \cos \omega_D T}{\sin \omega_D T} \tag{3.49}$$

This means that (3.48) transforms an analog low-pass $H(\omega_A)$ into a digital bandpass $H'(\omega_D T)$. Equation (3.48) can thus be used to design bandpass digital filters as follows:

1. Decide on a center frequency $\psi_0 T$† for the digital filter. This may be forced by the specifications, or it may be available to simplify the choice of other parameters. Compute the critical frequencies of the desired analog filter from the critical frequencies in the specification of the digital filter, using (3.49). Equation (3.49) will often yield negative frequencies, which is all right since $H(\omega_A)$ will be an even function of ω_A.
2. Design an analog filter $H(s)$ with the translated specifications. It is likely that one or more of the specifications will be superfluous.
3. Replace s by

$$s = \frac{z^2 - 2z \cos \psi_0 T + 1}{z^2 - 1}$$

† All that is ever really needed is $\cos \psi_0 T$.

in $H(s)$, and perform the required algebra to manipulate the resulting $H'(z)$ into a ratio of polynomials; this is the required filter.

One example will suffice to illustrate the method.

EXAMPLE

Design a digital bandpass filter for a 1,000-Hz sampling rate, to pass 100 to 400 Hz with ripple-free attenuation between 0 and 3 db. At 45 and 450 Hz, the filter must be at least 20 db down and must fall off monotonically beyond both frequencies.

1. $\psi_0 T$ can be chosen so that the two analog 3-db points are negatives of each other. If the digital 3-db points are L_1 and L_2, we can see from (3.49) that $\psi_0 T$ should be chosen so that

$$\cos \psi_0 T = \frac{\cos \frac{1}{2}(L_1 + L_2)}{\cos \frac{1}{2}(L_1 - L_2)} \tag{3.50}$$

Using (3.50) yields $\cos \psi_0 T = 0$ for this case. This means that the transformation from digital to analog critical frequencies becomes

$$\omega_A = -\cot \omega_D T$$

The 3-db points now translate to

$$\omega_{3db} = \pm 1.3764$$

The 20-db points are not equal in magnitude. They are

$$\omega_{s_1} = -3.442$$

$$\omega_{s_2} = 3.078$$

2. The problem is now to design a monotonic filter with 0 to 3 db in the region $0 < \omega < 1.3764$. The filter must be 20 db down by $\omega = 3.078$, which will automatically satisfy the requirement at $\omega = 3.442$. A Butterworth design seems to be called for, with $\omega_c = 1.3764$. For $\omega/\omega_c = 2.23$ we demand 20-db attenuation. We calculate the order n:

$$1 + (2.23)^{2n} \geq 100$$

It is clear that $n = 3$ will suffice. A unity-bandwidth Butterworth filter of order 3 is

$$H_1(s) = \frac{K}{s^3 + 2s^2 + 2s + 1}$$

$H(s)$ is found by replacing s by $s/1.3764$:

$$H(s) = \frac{2.6075581}{s^3 + 2.7528s^2 + 3.788954s + 2.6075581}$$

3. To find the required digital filter we let $s = (z^2 + 1)/(z^2 - 1)$ in $H(s)$. Note that, since this is going to yield a function of z^2, the resulting digital filter will be quite simple.

$$H'(z) = \frac{0.256919688(z^6 - 3z^4 + 3z^2 - 1)}{z^6 - 0.577263586z^4 + 0.421794133z^2 - 0.056299786}$$

If the gain of the filter is not important, this can be programmed with three multiplications per output point, plus a moderate number of additions.

The techniques of this section can also be used to design high-pass and band-elimination filters.

3.11 FREQUENCY-SAMPLING FILTERS

As a preliminary to discussion of the next filter type, we remind the reader that another way to approximate arbitrary functions is by means of sampling functions. Thus the function $f(t)$ can be approximated by

$$g(t) = \sum_{n=-\infty}^{\infty} f(nT) \frac{\sin (2\pi/T)(t - nT)}{(2\pi/T)(t - nT)} \quad \text{this is wrong} \quad \text{should be } \pi \qquad (3.51)$$

We know further that $g(t)$ is exactly equal to $f(t)$ if the latter is limited to a frequency band $1/2T$. Even if the band of $f(t)$ is not limited, $g(t)$ is exactly $f(t)$ at the sampling instants $t = nT$.

We inquire whether the sampling theorem serves as a useful basis for filter design. First it is necessary to show that an *elemental* filter can be constructed with the shape (as a function of frequency) of one of the interpolation functions of (3.51) and also the appropriate phase so that

$$= \sum f_{nT} \frac{\sin (\pi(t-n))}{\pi(t-n)}$$

$$\text{when } 0 \leq \tau \leq 1$$

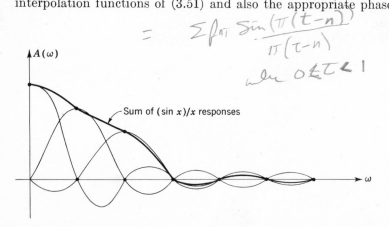

Fig. 3.17 Illustration of sampling.

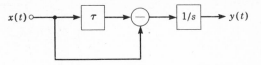

Fig. 3.18 Analog comb filter with zero canceled at origin.

the outputs of several elemental filters can be added as if they were scalars.   Then it must be ascertained whether a *finite* sum of the form (3.51) yields a suitable filter-selectivity curve.   For example, Fig. 3.17 shows a desired $|H(e^{j\omega T})|$ and three interpolation functions used to approximate $|H(e^{j\omega T})|$.

Consider now the configuration of Fig. 3.18.   The magnitude and phase of $H(s)$ are

$$|H(j\omega)| = \frac{2|\sin \omega \tau/2|}{\omega} \qquad \varphi(\omega) = -\frac{\omega \tau}{2}$$

These functions are shown in Fig. 3.19 and are seen to exhibit the properties of a low-pass filter with linear phase.

This argument may now be extended to the configuration of Fig. 3.20.

The amplitudes and phases at the outputs of each lossless resonator are shown in Fig. 3.21.   The fifth line of the figure indicates whether the two outputs are exactly in phase or exactly out of phase, and we see that only during the overlap intervals of the main lobes are they in phase opposition.   Therefore, by subtracting the two outputs, as indicated in Fig. 3.21, the amplitudes $A_i(\omega)$ are effectively *added* during overlap and subtracted otherwise.   This tends to create a composite amplitude $A(\omega)$

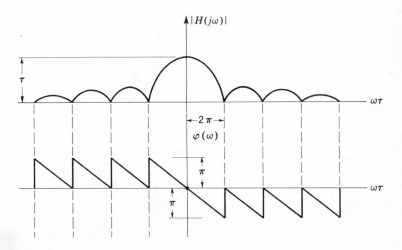

Fig. 3.19   Frequency response of filter of Fig. 3.18.

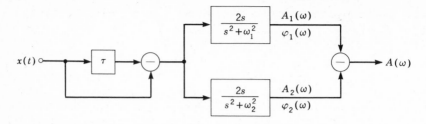

Fig. 3.20 Analog comb filter with zeros canceled by several resonators.

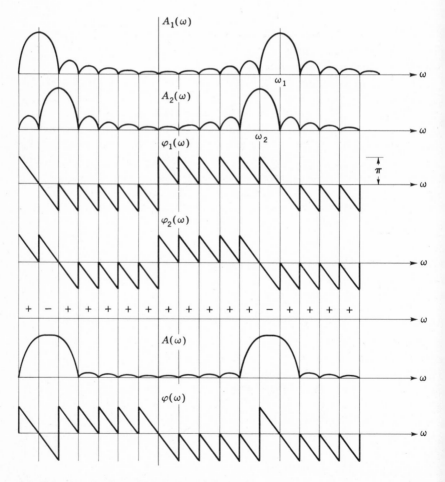

Fig. 3.21 Amplitude and phase functions for Fig. 3.20.

which is fairly constant during the overlap and which also tends to attenuate the side lobes. Furthermore, since the phases $\varphi_1(\omega)$ and $\varphi_2(\omega)$ are linear, the output phase $\varphi(\omega)$ must also be linear with frequency. Extending this argument to the case of n lossless resonators, we can achieve an effect identical to that produced by adding true interpolation functions of the type shown in Fig. 3.17. Thus a general bandpass filter with linear phase over the entire frequency band can be specified by *sampling* a desired magnitude-vs.-frequency curve at selected, uniformly spaced frequencies.

A practical disadvantage of this method for analog-filter design is the need for lossless resonators. To a large extent, this difficulty is overcome in applying the technique to digital-filter design.

.12 TECHNIQUE 4: FREQUENCY-SAMPLING TECHNIQUE

The difference equation

$$y(nT) = x(nT) - x(nT - mT) \tag{3.52}$$

has the z transfer function $1 - z^{-m}$, which has m zeros equally spaced around the unit circle, at points

$$z_k = e^{j2\pi k/m} \qquad k = 0, 1, \ldots, m - 1 \tag{3.53}$$

If, in (3.53), the subtraction is replaced by an addition, the transfer function becomes $1 + z^{-m}$, for which the zeros are also equally spaced around the unit circle, at

$$z_k = \exp\left(\frac{j2\pi(k + 1/2)}{m}\right) \qquad k = 0, 1, \ldots, m - 1 \tag{3.54}$$

The magnitude-vs.-frequency curves of these filters repeat with period $2\pi/m$ radians; these filters are comb filters and can be incorporated into an especially appealing type of digital design.

Before presenting this design, let us examine the practical significance of (3.52). If successive values of $x(nT)$ are thought of as contents of registers in computer memory, it is clear that m registers must be set aside for buffer storage in order to do the computation specified by (3.52). This is usually a substantial amount of memory compared with that needed for computation of second-order difference equations. Thus possible practical use of the filters we are about to describe is limited to systems where the necessary memory is available or to systems where many filters share a common input. It is important to note that the delays implied by (3.52) can be effected by digital delay lines, which are presently relatively cheap forms of memory.

A simple resonator can be placed in cascade with the comb filter.

Let the resonator be comprised of two complex conjugate poles which lie directly on the unit circle. The discussion of the zeros of the resonator will be deferred. Further, let the angle $\omega_r T$ of a resonator pole be such that the pole is coincident with a zero of the comb filter.

$$\omega_r T = \left\{ \begin{array}{c} \dfrac{2\pi k}{m} \\ \dfrac{2\pi(k + 1/2)}{m} \end{array} \right\} \text{ for a comb filter of the } \left\{ \begin{array}{c} \text{first} \\ \text{second} \end{array} \right\} \text{ type}$$

The poles of the resonator cancel the kth zero of the comb filter and its conjugate. In what follows, we shall refer to the resonator used to cancel the kth zero as the kth elemental filter.

1. The impulse response is of finite duration, mT.
2. The magnitude-vs.-frequency response is

$$|H(e^{j\omega T})| = \left| \frac{\sin (m\omega T/2)}{\cos \omega T - \cos \omega_r T} \right|$$

which is zero at all the radian frequencies for which the comb filter is zero, except at $\omega_r T$. The magnitude at $\omega_r T$ is $(m/2) \csc \omega_r T$.
3. The phase vs. frequency is exactly linear except for discontinuities of π radians. These discontinuities occur where the magnitude response is zero.
4. The phase difference between two composite filters with resonant frequencies $\omega_r T$ and $\omega_{r+1} T$ is π for $\omega_r < \omega < \omega_{r+1}$ and is zero outside these bounds.
5. The amplitude of any composite filter is zero at the resonant frequencies of all the other composite filters.

As m is made large, the magnitude response of a cascaded filter becomes like

$$\left[\frac{\sin (\omega - \omega_r) T}{(\omega - \omega_r) T} + \frac{\sin (\omega + \omega_r) T}{(\omega + \omega_r) T} \right]$$

in shape. These properties suggest that any desired magnitude response could be obtained by adding together the weighted outputs of cascaded comb and elemental filters, just as any "band-limited" time function can be formed from a weighted sum of delayed $(\sin t)/t$ functions. Let us examine this idea, which we call frequency sampling, in some detail.

A sufficiently *narrowband* frequency-response function (one for which the frequency response is a sufficiently smooth function of frequency) is sampled at equally spaced points, with radian frequencies

$$\omega_k T = \frac{2\pi k}{m} \quad \text{or} \quad \frac{2\pi(k + 1/2)}{m} \quad k = 0, 1, \ldots, m - 1$$

depending on which kind of comb filter will be used. Let the sample value of the amplitude at frequency $\omega_k T$ be W_k. An elemental filter of resonant frequency $\omega_k T$ cascaded with a comb filter of delay mT and a gain of $W_k \sin \omega_k T$ is used to provide an *elemental frequency response* of W_k at radian frequency $\omega_k T$ and zero at the other sampling frequencies.

Since the phases at resonance of the consecutive elemental filters differ by π, the gains of the odd-numbered elemental filters are to be multiplied by -1. The desired input to the filter is applied to the comb filter, which is shared among all the elemental filters, followed by the gains (and sign changes for odd-numbered elemental filters). The outputs of all the elemental filters, with proper gains, are added together to give the desired filter output. The resulting filter has an impulse response of duration nT, a frequency response with linear phase, and an amplitude response that agrees with specifications at the sampling frequencies and connects the sampling points smoothly. The variety of filters that can be programmed with this technique is fairly large.

There are some practical problems to be considered before the frequency-sampling method is applied. For one thing, the resonant poles of an elemental filter cannot exactly cancel the zeros of a comb filter because of quantization. Thus it is wise to move both the zeros of the comb filter and the poles of the elemental filters slightly inside the unit circle, with a radius of something like $e^{-aT} = 1 - 2^{-26}$.

We have successfully programmed filters with the poles and zeros at radii ranging from $1 - 2^{-12}$ to $1 - 2^{-27}$, with little change in the behavior of the filter.

Let us now turn to the zeros of the resonators, which are especially important in the design of bandpass filters. In the passband it is common for the samples W_k to be equal. Thus it would be desirable if all the elemental filters had the same gains at resonance. If a zero is put at $(\cos \omega_k T)e^{-aT}$, the gain of each elemental in cascade with the comb filter becomes $m/2$. This has a slight effect on the magnitude and phase response, negligible for large m. The modified comb-filter z-transform thus becomes

$$H(z) = 1 - e^{-maT}z^{-m} \tag{3.55}$$

and the modified kth elemental filter becomes

$$H_k(z) = \frac{1 - e^{-aT} \cos \omega_k T z^{-1}}{1 - 2e^{-aT} \cos \omega_k T z^{-1} + e^{-2aT}z^{-2}} \tag{3.56}$$

It is worth noting that the introduction of the additional zero does not require another multiplication since twice the numerator coefficient is also present in the denominator. The response of a filter of the form of (3.55) in cascade with (3.56) is shown in Fig. 3.22. As shown in Sec.

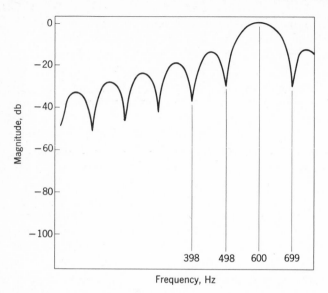

Fig. 3.22   Measured frequency response of comb filter in series with digital resonator.

3.6, case 1, even more uniform frequency response would be obtained by deleting e^{-aT} from the numerator of (3.56).

EXAMPLE: BANK OF BANDPASS FILTERS

It was desired to design a bank of bandpass filters (with a common input), each 400 cps wide, covering the band 300 to 3,100 Hz.   The filters were to be as selective as possible but with minimum ringing time.   A further requirement was that the contiguous filters cross at -3 db of the midband gain.   None of the standard designs had satisfactory selectivity combined with short ringing.

The filters chosen were frequency-sampling filters, each composed of seven elemental filters.   The consecutive zeros were 100 Hz apart.   Since the sampling rate was 12.5 kHz, m was 125.   The general form of such a filter has a z-transform

$$H(z) = (1 - e^{-maT}z^{-m})$$

$$\times \sum_{k=r}^{r+6} (-1)^k \frac{1 - e^{-aT}\cos{(2\pi k/m)}z^{-1}}{1 - 2e^{-aT}\cos{(2\pi k/m)}z^{-1} + e^{-2aT}z^{-2}} W_k \quad (3.57)$$

The design of the filter is completely specified by choosing r and the set of W_k in (3.57).   Since 3-db crossovers 400 Hz apart were required, W_{r+1} and W_{r+5} were chosen to be 0.707.   The three center terms had

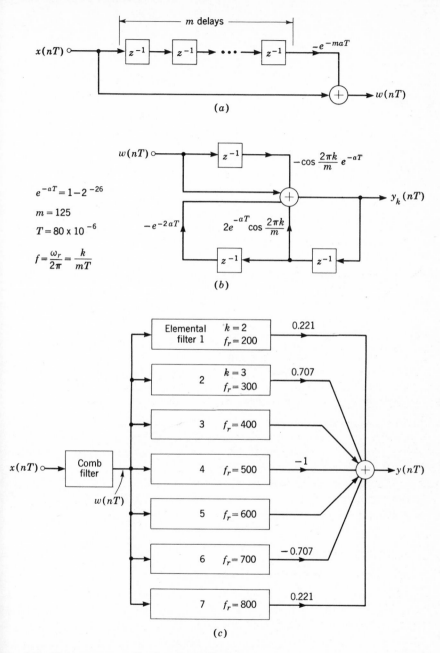

$$e^{-aT} = 1 - 2^{-26}$$

$$m = 125$$

$$T = 80 \times 10^{-6}$$

$$f = \frac{\omega_r}{2\pi} = \frac{k}{mT}$$

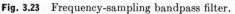

Fig. 3.23   Frequency-sampling bandpass filter.

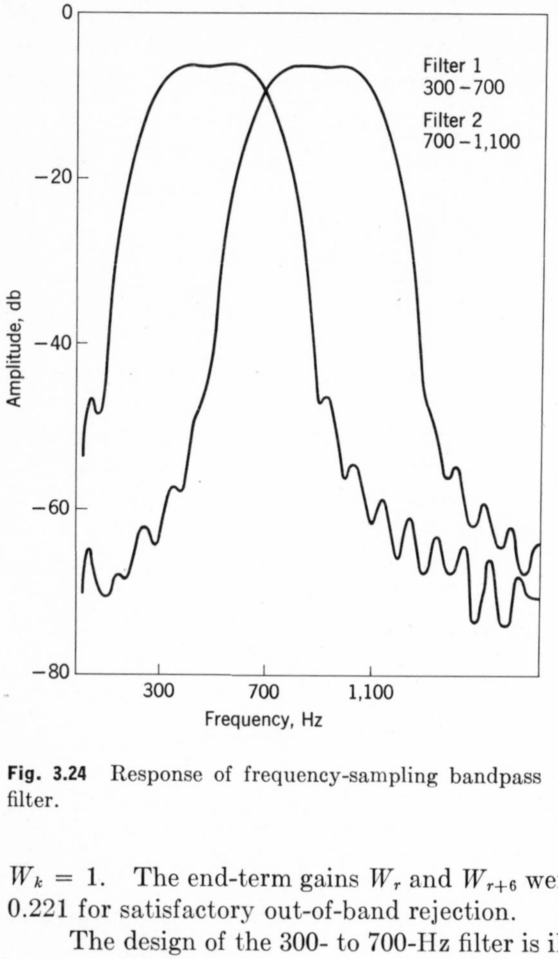

Fig. 3.24 Response of frequency-sampling bandpass filter.

$W_k = 1$. The end-term gains W_r and W_{r+6} were found empirically to be 0.221 for satisfactory out-of-band rejection.

The design of the 300- to 700-Hz filter is illustrated in Fig. 3.23, and the experimental frequency responses of it and the next higher filter (700 to 1,100 Hz) are shown in Fig. 3.24.

3.13 DESIGN EXAMPLES

1. PROGRAM TO DESIGN DIGITAL ELLIPTIC FILTERS

In this example, we briefly describe the computational details of a computer program that will compute the poles and zeros of an elliptic digital low-pass filter, given the specifications. The formulas used in this section are based on the theory of Sec. 3.8. We have chosen the bilinear transformation as the one relating the analog and digital elliptic filters; this implies, following the discussion of Sec. 3.10, that the cutoff radian frequency ω_c is to be replaced by $\tan (\omega_c T/2)$ and ω_r is to be replaced by

$\tan(\omega_r T/2)$.   Assume we desire a passband ripple of $\frac{1}{2}$ db and a ripple magnitude in the stopband of -60 db relative to the zero frequency magnitude.   The cutoff radian frequency ω_c is specified as $2\pi \times 2,000$ rad / sec and the beginning of the stopband $\omega_r = 2\pi \times 2,200$ rad/sec.   For Fig. 3.10, this means that $A = 1,000$ and $\epsilon^2 = 0.122$, $\epsilon = 0.349285$.   We choose $T = 10^{-4}$ sec.

1. Find k and k_1.†

$$k_1 = \frac{\epsilon}{\sqrt{A^2 - 1}} = 3.4928514 \times 10^{-4}$$

$$k = \frac{\tan(\omega_c T/2)}{\tan(\omega_r T/2)} = 0.87823911$$

2. Compute the complete elliptic integrals $K(k)$, $K(k_1)$, $K'(k)$, $K'(k_1)$.
   An algorithm for computing complete elliptic integrals is given by Hastings [6].   Tables of these functions can be found in Abramowitz [1].   In our example,

   $K(k) = 2.1965323 \qquad K(k_1) = 1.5707964$

   $K'(k) = 1.674388 \qquad K'(k_1) = 9.3459166$

3. To find the order N of the required filter, we use formula (3.27),

   $$N = \frac{K'(k_1)K(k)}{K(k_1)K'(k)}$$

   which gives $N = 7.80519$.   Since N must be an integer, we choose $N = 8$.   In most applications, k is close to unity and k_1 is much smaller than unity; in such cases, the required N can be computed from the approximation

   $$N \approx \frac{2}{\pi^2} \ln \frac{4A}{\epsilon} \ln \frac{8 \tan(\omega_c T/2)}{\tan(\omega_r T/2) - \tan(\omega_c T/2)}$$

4. For $k \approx 1$ and $k_1 \ll 1$, the following approximation holds for the real component u_0 of the poles in the λ plane:

   $$u_0 \approx -\frac{K'(k)}{K'(k_1)} \ln(\sec \varphi + \tan \varphi)$$

   $$= -\frac{K'(k)}{K'(k_1)} \ln \frac{\sqrt{(1 + \epsilon^2)} + 1}{\epsilon} \qquad \text{with } \tan \varphi = 1/\epsilon$$

   $u_0 = -0.31786181$

† The values obtained for elliptic functions for this and the following example are approximations, and should not be considered accurate to the number of decimal places given.

5. Given N and u_0, the poles and zeros in the λ plane can be instantly set down as shown in Fig. 3.12, which holds for N even. To find the corresponding z-plane poles and zeros requires the transformations given by (3.28) followed by the bilinear transformation given by (3.35). The transformation (3.28) is carried out via (3.29) where predistortion of the normalized analog filter is accomplished by replacing ω_c in (3.29) by $\tan (\omega_c T/2) = 0.7265425$. Figure 3.25

Poles	Zeros
$0.58921995 \pm j0.2763795$	$-0.80925294 \pm j0.5874599$
$0.4509549 \pm j0.67726114$	$-0.17915135 \pm j0.98382143$
$0.3420187 \pm j0.8611878$	$0.09481303 \pm j0.99549506$
$0.29985251 \pm j0.93454875$	$0.178863 \pm j0.983874$

shows a sketch of these poles and zeros in the z plane and Fig. 3.26 shows the corresponding frequency-response magnitude function. It is seen that the original specifications have been met.

All the computations implied by the above steps are routine and can easily be programmed on a general-purpose computer, with the possible exception of the elliptic functions required for (3.28). For this, we first

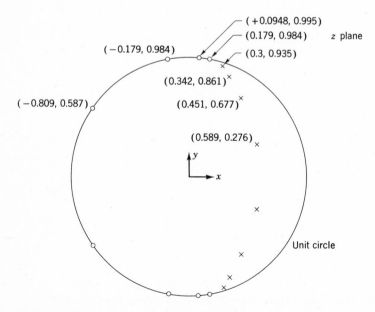

Fig. 3.25 Poles and zeros of digital elliptic filter.

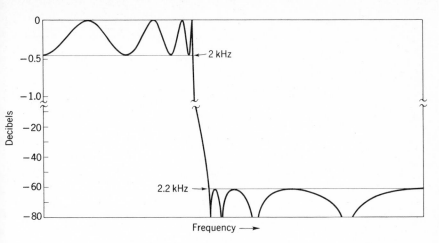

Fig. 3.26 Computed response of digital elliptic filter.

note the relations (3.29). Since cn u and dn u are easily found from sn u, only an algorithm for sn u is needed; this can be obtained from the expression of sn u in terms of theta functions, as follows:

$$\operatorname{sn}(y,k) = \frac{1}{\sqrt{k}} \frac{\theta_1(y/2K,q)}{\theta_4(y/2K,q)} \tag{3.58}$$

where the theta functions are defined by the infinite series

$$\theta_4\left(\frac{y}{2K}, q\right) = 1 + 2\left(-q \cos\frac{2\pi y}{2K}\right.$$

$$\left. + q^4 \cos\frac{4\pi y}{2K} - q^9 \cos\frac{6\pi y}{2K} + \cdots\right)$$

$$= 1 + 2\sum_{n=1}^{\infty} (-1)^n q^{n^2} \cos\frac{2n\pi y}{2K} \tag{3.59}$$

$$\theta_1\left(\frac{y}{2K}, q\right) = 2q^{1/4} \sin\frac{\pi y}{2K} - 2q^{9/4} \sin\frac{3\pi y}{2K} + 2q^{25/4} \sin\frac{5\pi y}{2K} \cdots$$

$$= 2\sum_{n=0}^{\infty} (-1)^n q^{(n+1/2)^2} \sin(2n+1)\frac{\pi y}{2K}$$

The parameter q is given by

$$q = e^{-\pi K'(k)/K(k)} \tag{3.60}$$

In the above expressions the variable y is real and the Jacobian elliptic function is also real. Since q is generally rather small, the series (3.59) converges very rapidly and can be used to define the algorithm for

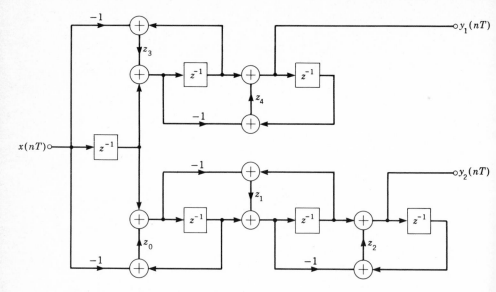

Fig. 3.27   An all-pass phase splitter.

the computer program.   Even if q is rather close to unity, convergence is still fairly rapid.

2. DESIGN OF A 90° ELLIPTIC DIGITAL PHASE-SPLITTING NETWORK

It also is possible to use mappings based on the Jacobian elliptic functions together with the bilinear transform to design a pair of all-pass networks whose phase difference is the closest possible approximation to 90° over an interval of frequencies.   It is worth commenting that any digital network is all pass provided that all the poles and zeros can be put in reciprocal pairs, e.g.,

$$H(z) = \prod_i \frac{(a_i - z^{-1})}{(1 - a_i z^{-1})} \tag{3.61}$$

It is possible to arrange such a network so that only one multiplication is necessary for each pole-zero pair.   Figure 3.27 shows a digital network with two outputs, $y_1(nT)$ and $y_2(nT)$, such that each is all pass relative to the input.   It should not be difficult for the reader to verify that the output $y_1(nT)$ sees a transfer function with poles z_3 and z_4 and with zeros $1/z_3$ and $1/z_4$, while the transfer function associated with $y_2(nT)$ has poles z_0, z_1, and z_2 and zeros $1/z_0$, $1/z_1$, and $1/z_2$.   The optimum digital network of this sort, as a 90° phase splitter, would have a characteristic as shown

n Fig. 3.28.   This is, of course, only a phase difference.   Neither network has, by itself, a particularly useful phase characteristic.

The design of such networks is comparable in its intricacy to the
design of equiripple low-pass filters, and we will confine ourselves here to
giving a "cookbook" procedure for approximating 90° with an error of
± ε in the frequency range $\theta_a \leq \omega T \leq \theta_b$.   We shall accompany the procedure with a specific example in which the permitted error is 1° in the
range from 10° to 120°.

1. Compute

$$k = \frac{\tan (\theta_a/2)}{\tan (\theta_b/2)} = 8.74887 \times 10^{-2}$$

$$k_1 = \left(\frac{1 - \tan (\epsilon/2)}{1 + \tan (\epsilon/2)}\right)^2 = 0.96569$$

2. Compute

$$N = \frac{K'(k)K(k_1)}{K'(k_1)K(k)} \approx 4.8$$

and force N to be the next higher integer; therefore in our example,
$N = 5$.

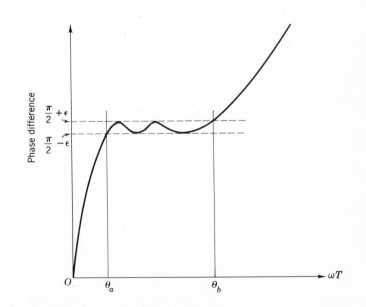

Fig. 3.28   Response of a 90° phase splitter.

3. Compute the quantities

$$p_\ell = - \tan \left(\frac{\theta_a}{2}\right) \frac{\text{sn}\,[(4\ell + 1)K'(k)/2N,k]}{\text{cn}\,[(4\ell + 1)K'(k)/2N,k]}$$

$$\text{for } \ell = 0, 1, \ldots, N - 1$$

$p_0 = -0.03950304$

$p_1 = -0.3892747$

$p_2 = -3.8360295$

$p_3 = 1.0039420$

$p_4 = 0.1509398$

4. For those p_ℓ which are negative compute the network coefficients

$$z_\ell = \frac{1 + p_\ell}{1 - p_\ell}$$

$z_0 = 0.923997$

$z_1 = 0.4396001$

$z_2 = -0.5864377$

and for those p_ℓ which are positive compute the network coefficients

$$z_\ell = \frac{1 - p_\ell}{1 + p_\ell}$$

$z_3 = -0.0019671$

$z_4 = 0.7377101$

The positive p_ℓ give the poles of one branch of the network while the negative p_ℓ give the poles of the other branch (by a different formula). This design procedure is modeled after that in Storer [14] but extended to the case where the total number of poles in both networks may be odd or even.

3.14 SUMMARY

The theoretical background was presented for the frequency-domain design of digital filters. We took the point of view that filter theory is a general theory that can be applied to the construction of either analog or digital filters. It was first established that, for any realizable analog filter with a rational function as transfer function, a digital filter can be found with the same impulse response (at the sample values) as that of the analog filter. For narrowband filters where aliasing problems are minor, this means that the frequency response of the digital filter will also be close to the frequency response of the analog filter. This technique was referred to as impulse-invariant. For wideband filters, aliasing may make this technique unfeasible. If the filter specification is an approximation to an ideal filter whose frequency response consists of connected constant segments, then a variety of approaches, often leading to the same design, can be applied. For example, by using the potential analogy approach, the poles and zeros of a filter can be placed in a λ plane in such a way that the desired frequency response is obtained along some curve in the λ plane. If, then, one can succeed in mapping the λ plane into the z plane so that that curve maps into the unit circle in the z plane and if the mapped z-plane poles lie within the unit circle, the desired filter can be derived. Similarly, knowing that the transfer function of a digital filter can be expressed as a rational function of powers of trigonometric functions in ωT allows for the specification of a digital filter via a desired squared-magnitude function. Sometimes these designs can be conveniently realized by applying the bilinear transformation directly to an already designed analog filter.

The above-mentioned techniques deal with filters that require both poles and zeros for their realization and therefore have impulse responses of theoretically infinite duration. If a filter is defined by a finite-duration impulse response, then a sampling theorem holds in the frequency domain, so that the complex frequency response can be represented by a finite number of elemental frequency-response functions. This makes it feasible to design a filter by specifying the response at a finite number of equally spaced frequencies with the assurance that the actual response will be precisely as specified at these frequencies. In this way, bandpass filters with sharp skirts and linear phase can be designed. These digital filters can be realized by using comb filters and lossless digital resonators.

APPENDIX: REALIZABILITY OF DIGITAL FILTERS
WITH CERTAIN SQUARED-MAGNITUDE FUNCTIONS

In the following we shall find it useful to define polynomials of the form

$$a_0 z^k + a_1 z^{k-1} + \cdots + a_1 z + a_0$$

such that the coefficient of z^r is equal to the coefficient of z^{k-r}, as mirror-image polynomials of order k, and to summarize some of the properties of the polynomials.

1. A mirror-image polynomial (MIP) of order k has roots that occur in reciprocal pairs for k even.

Proof: Let z be replaced by $1/z$ in the original MIP and equate it to zero.

$$a_0 z^{-k} + a_1 z^{-k+1} + a_1 z^{-1} + a_0 = 0$$

If we multiply by z^k, which introduces no new roots to the equation, we see that the roots of the polynomial in $1/z$ are the same as the roots of the polynomial in z.   Thus the roots either occur in reciprocal pairs or are self-reciprocals.   For k odd, one of the roots cannot be part of a reciprocal pair.   This root is its own reciprocal, and since it must be real, it is either $+1$ or -1.   We shall generally not be interested in odd-order MIPs.

2. The sum of mirror-image polynomials of the same order is an MIP of that order.
3. A mirror-image polynomial of order k plus z^r times a mirror-image polynomial of order $k - 2r$ is a mirror-image polynomial of order k.
4. The product of mirror-image polynomials is a mirror-image polynomial.
5. A polynomial whose roots occur in reciprocal pairs is an MIP.   This property is proved by multiplying together the factors that are reciprocals, noting that each product is an MIP and applying statement 4.

We shall use the above properties of mirror-image polynomials to prove the existence of digital filters with certain squared-magnitude functions.   Suppose a squared-magnitude function $F(\omega T)$ is given.   We can always replace ωT by $-j \ln z$, equivalent to $z = e^{j\omega T}$, to give a function $G(z)$ which is equal to $F(\omega T)$ when evaluated along the unit circle.   For our purposes, $G(z)$ must be a rational function of z with real coefficients. We can guarantee this if, in the squared-magnitude function, ωT appears as only part of the following expressions: $\sin^2 (\omega T/2)$, $\cos^2 (\omega T/2)$, $\tan^2 (\omega T/2)$, $\cot^2 (\omega T/2)$, $\sec^2 (\omega T/2)$, $\csc^2 (\omega T/2)$, and if these expressions are combined only by addition, subtraction, multiplication, and division.   This is because the squared trigonometric functions of $\omega T/2$ yield rational functions of z, and rational functions form a field with multiplication and addition.   The correspondence of squared trigonometric functions of $\omega T/2$ with rational functions of z is given in Table 3.1.

Now let us assume that we have been given an $F(\omega T)$ with the preceding restrictions, found $G(z)$, and found all the poles and zeros of $G(z)$. To find a transfer function $H(z)$ with squared-magnitude function $F(\omega T)$, we should like to use the following procedure:

1. Discard any poles or zeros at the origin.
2. Put the remaining poles in one-to-one correspondence with each other in such a way that for each pair of poles the distance from one pole

Table 3.1

$$\sin^2 \frac{\omega T}{2} \rightarrow \frac{(z-1)^2}{-4z}$$

$$\cos^2 \frac{\omega T}{2} \rightarrow \frac{(z+1)^2}{4z}$$

$$\tan^2 \frac{\omega T}{2} \rightarrow \frac{-(z-1)^2}{(z+1)^2}$$

$$\cot^2 \frac{\omega T}{2} \rightarrow \frac{-(z+1)^2}{(z-1)^2}$$

$$\sec^2 \frac{\omega T}{2} \rightarrow \frac{4z}{(z+1)^2}$$

$$\csc^2 \frac{\omega T}{2} \rightarrow \frac{-4z}{(z-1)^2}$$

to any point on the unit circle is in constant ratio to the distance
from that point on the unit circle to the other pole of the pair.

3. Repeat step 2 for the zeros.

4. Discard one of each corresponding pair of poles and zeros, making sure
that the retained poles are all within the unit circle.

5. Form $H(z)$ from the retained poles and zeros only.

This procedure works only if steps 2 and 3 are possible. The magnitude of $H(z)$ is proportional to the product of the distances from the unit
circle to retained zeros divided by the product of distances to the retained
poles and thus is proportional to the square root of $F(\omega T)$. Note that
the singularities of $G(z)$ at the origin contribute nothing to a magnitude
function since the distance from the origin to the unit circle is always 1.

Steps 2 and 3 are possible if the poles (and zeros) occur in pairs
whose locations are equal or, more commonly, conjugate reciprocals of
each other, $re^{j\theta}$ and $(1/r)e^{j\theta}$. Since the singularities of $G(z)$ occur in conjugate pairs it is enough to require the roots to occur in equal or reciprocal
pairs. By the following proof we show that the ratio of distances from
any point on the unit circle to two singularities that are conjugate reciprocals is constant.

Proof: Let the two fixed singularities be $re^{j\theta}$ and $(1/r)e^{j\theta}$. By the
theorem of Apollonius, the locus of points the ratio of whose distance to
two fixed points is constant is a circle whose center is colinear with the
two fixed points. Consider the ratios at the two points $e^{j\theta}$ and $e^{j(\theta+\pi)}$.
The ratios are equal to r at both points, and the only circle passing through
both points whose center is colinear with the two fixed points is the unit
circle. Thus the unit circle is the circle of Apollonius, and the ratio of
the distance to the singularities is constant.

The corresponding proof for equal singularities is trivial.

If, after poles or zeros at the origin are discarded, the numerator and denominator of $G(z)$ become mirror-image polynomials, it is possible by following the five-step procedure given above to find $H(z)$ with squared-magnitude function $F(\omega T)$.† We shall now enumerate a few types of $F(\omega T)$ which yield mirror-image polynomials for numerator and denominator of $G(z)$.

Consider $F(\omega T)$ a rational function of $\sin^2 (\omega T/2)$. Then $G(z)$, by substitution from Table 3.1, is of the form

$$G(z) = \frac{a_0 \left[\dfrac{(z-1)^2}{-4z}\right]^n + a_1 \left[\dfrac{(z-1)^2}{-4z}\right]^{n-1} + \cdots}{b_0 \left[\dfrac{(z-1)^2}{-4z}\right]^m + b_1 \left[\dfrac{(z-1)^2}{-4z}\right]^{m-1} + \cdots}$$

which can be put in the form

$$G(z) = \frac{\{a_0[(z-1)^2]^n + a_1[(z-1)^2]^{n-1}(-4z) + a_2[(z-1)^2]^{n-2}(-4z)^2 + \cdots\}(-4z)^{m-n}}{b_0[(z-1)^2]^m + b_1[(z-1)^2]^{m-1}(-4z) + b_2[(z-1)^2]^{m-2}(-4z)^2 + \cdots}$$

The $(-4z)^{m-n}$ contributes only poles or zeros at the origin and is discarded. Note that powers of $(z - 1)^2$ are mirror-image polynomials by property 5, and thus the entire numerator of $G(z)$ is a mirror-image polynomial, by property 3. The same reasoning holds for the denominator.

Exactly similar reasoning shows that rational functions of $\cos^2 (\omega T/2)$ also yield realizable digital filters. Rational functions of $\tan^2 (\omega T/2)$ yield $G(z)$ of the form

$$G(z) = \frac{\{a_0(z-1)^{2n} + a_1(z-1)^{2(n-1)}(z+1)^2 + \cdots + a_n(z+1)^{2n}\}(z+1)^{2(m-n)}}{b_0(z-1)^{2m} + b_1(z-1)^{2(m-1)}(z+1)^2 + \cdots + b_m(z+1)^{2m}}$$

Here we note by repeated application of property 4 that again both numerator and denominator are mirror-image polynomials and therefore a digital filter exists with the desired squared-magnitude function. Again, exactly similar reasoning suffices to extend the proof to rational functions of $\cot^2 (\omega T/2)$, $\sec^2 (\omega T/2)$, and $\csc^2 (\omega T/2)$.

Similar discussions show that rational functions of sums of products or ratios of the squares of the trigonometric functions of Table 3.1, that is, $\sin^2 (\omega T/2) + \tan^4 (\omega T/2)$, also yield $G(z)$, which is a ratio of mirror-

† This is not strictly correct if the mirror-image polynomials have simple roots on the unit circle, which are then their own conjugate reciprocals. Subsequent statements should be qualified by taking this possibility into account. However, this is a limiting case, unlikely to occur.

image polynomials. Further extensions can be made, but we shall be content here with these few examples and techniques.

REFERENCES

1. Abramowitz, M., and I. A. Stegun: "Handbook of Mathematical Functions," Dover Publications, Inc., New York, 1965.
2. Craig, J.: unpublished notes, 1963.
3. Golden, R. M.: Digital Computer Simulation of a Sampled Data Voice-excited Vocoder, *J. Acoust. Soc. Am.*, **35**: 1358–1366 (1963).
4. Golden, R. M., and J. F. Kaiser: Design of Wideband Sampled Data Filters, *Bell System Tech. J.*, **43**(4): (July, 1964).
5. Guillemin, E. A.: "Synthesis of Passive Networks," John Wiley & Sons, Inc., New York, 1957.
6. Hastings, C., Jr.: "Approximation for Digital Computers," Princeton University Press, Princeton, N.J., 1955.
7. Jacobi, G. J.: "Fundamenta Nova Theoriae Functionum Elliptecarium," Konigsberg, 1829.
8. James, H. M., N. B. Nichols, and R. S. Phillips: "Theory of Servomechanisms," chap. 5, pp. 231–261, McGraw-Hill Book Company, New York, 1947.
9. Jury, E. I.: "Theory and Application of the z-Transform Method," John Wiley & Sons, Inc., New York, 1964.
10. Lerner, R. M.: Band-pass Filters with Linear Phase, *Proc. IEEE*, **52**: 249–268 (1964).
11. Lewis, M.: Synthesis of Sampled Signal Networks, *IRE Trans. Circuit Theory*, March, 1960.
12. Rader, C., and B. Gold: Digital Filter Design Techniques in the Frequency Domain, *Proc. IEEE*, **55**: 149–171 (1967).
13. Ragazzini, J. R., and G. F. Franklin: "Sampled-data Control Systems," McGraw-Hill Book Company, New York, 1958.
14. Storer, J. E.: "Passive Network Synthesis," McGraw-Hill Book Company, New York, 1957.
15. Weinberg, L.: "Network Analysis and Synthesis," McGraw-Hill Book Company, New York, 1962.
16. White, W. D., and A. E. Ruvin: Recent Advances in the Synthesis of Comb Filters, *IRE Natl. Conv. Record*, **5**: 186–199 (1957).
17. Whittaker, E. T., and G. N. Watson: "Modern Analysis," Cambridge University Press, New York, 1902.

4
Quantization Effects
in Digital Filters

4.1 GENERAL DISCUSSION

Thus far our analysis has been based on the mathematical theory of linear difference equations with constant coefficients. Thus, implicitly, we have assumed that the constant and variable parameters used in the generation of the equations are continuous; i.e., they can take on any value. In an actual realization of a digital filter, all these parameters are really discrete, since the word length of any digital device is finite. The introduction of this discretization into the filter equations results in nonlinear relationships which, except for a few very simple cases, are difficult, perhaps impossible, to deal with rigorously. Luckily, when the amount of quantization is small compared with the values of the signals and the parameters, the theory becomes much simpler, and many useful results can be obtained which are verifiable experimentally. The treatment in this chapter is primarily a development of the approximate model based on relatively small quantization errors, for fixed-point computation.

To illustrate the problem, we consider the simple equation

$$y(nT) = Ky(nT - T) + x(nT) \tag{4.1}$$

The quantity K in (4.1) is a constant, presumably filed in a register in the computer memory.   Thus, K can assume only certain discrete values and, in general, can only approximate the chosen value.   The effect here is the same as one encounters in the design of analog filters; the design of a specified filter might lead to an inductance of 10.2976 henrys but when the coil is wound and the inductance carefully measured, it turns out to be 10.331 henrys.   Since we can attain neither exact inductance values nor exact values of K, the problem becomes one of searching for filter realizations that are less sensitive to these fixed parameter errors [9, 13, 14].

A qualitatively different effect is caused by quantization of the input signal, $x(nT)$, in (4.1).   Whether or not the input to a digital filter is considered to be quantized depends on the situation.   If the input is inherently discrete, no error exists.   In a great many practical cases the input signals are inherently continuous, and analog-to-digital conversion is necessary before digital processing can be done.   Thus there is a basic source of error in this conversion.   A-D converters have been built to yield as many as 15 bits.   Figure 4.1 shows the action of a 15-level A-D converter, with constant level differences E_0.

The analog-to-digital converter of Fig. 4.1 effectively quantizes the signal.   We can distinguish several kinds of quantization.   The kind we shall call *rounding*, illustrated by Fig. 4.1, approximates the signal by the nearest quantization level.   The kind we shall call *truncation* approximates the signal by the highest quantization level that is not greater than the signal (and is equivalent to rounding the signal less one-half a quantization step).   The kind we shall call *sign-magnitude truncation* is like truncation for positive signals, but negative signals are approximated by the nearest quantization level not less than the signal.   We assume the

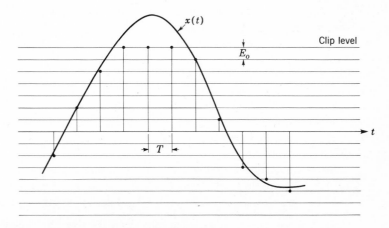

Fig. 4.1   Linear quantization of analog signal.

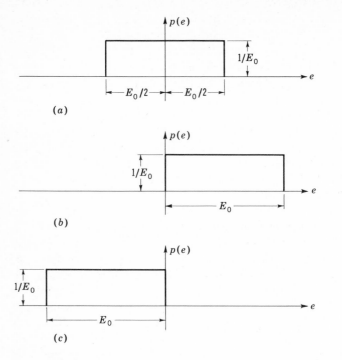

Fig. 4.2 Probability densities for roundoff and truncation.

error $e(nT)$ associated with the samples to be uniformly distributed. In the case of rounding, we expect the distribution shown in Fig. 4.2a. For truncation we expect the distribution shown in Fig. 4.2b. For sign-magnitude truncation we expect the distribution shown in Fig. 4.2b or c, depending on whether the signal is positive or negative.

If the signal fluctuation is such that many quantization levels can be traversed from one sample to the next, it seems reasonable to expect that the error $e(nT)$ at any sampling time will be statistically independent of $e(mT)$, the error at any other sampling time.† It is easy to give contrary examples (for instance, when the signal is constant); however, Bennett [1] has shown that this assumption holds for nearly all signals likely to be encountered in practice. In the theoretical developments of this chapter, we require a somewhat weaker assumption, namely, that noise samples be uncorrelated.

Given quantization errors, each with the probability density shown in Fig. 4.2, it is apparent that the effect is that of noise superimposed on the original analog signal. The input in (4.1) can thus be expressed as

† This assumption is for *rounding*. For truncation the error has a dc component. For sign-magnitude truncation the error is highly correlated with the signal.

$x(nT) = x_0(nT) + e(nT)$, where $x_0(nT)$ can be thought of as a noiseless input and $e(nT)$ is the added noise.   From Fig. 4.2 we can quickly compute the variance of the noise as

$$\sigma_e^2 = \frac{E_0^2}{12} \tag{4.2}$$

for the case of rounding or truncation.

The behavior of a linear discrete system such as that of (4.1) in the presence of added noise can readily be computed by straightforward techniques.   Before doing this, however, we shall introduce the third, and perhaps most interesting, source of quantization error, namely, the error introduced by quantizing the results of the multiplications needed to execute the iteration.   In (4.1), the results obtained when K and $y(nT - T)$ are multiplied must be quantized.   If K and $y(nT - T)$ each contain ν bits, their product will contain 2ν bits.   If the length, in bits, of this product is not reduced, then, for each successive iteration it will grow by ν bits.   Roundoff is thus seen to be unavoidable.   Its effect on the filter is not so easy to perceive as is the additive effect of A-D conversion of an analog input and, in fact, is strongly dependent on exactly how the quantization is done and on the particular realization of the digital filter. If, however, we assume that this roundoff error is independent from one iteration to the next, this leads to a model wherein noise sources are introduced at various nodes in the digital network rather than simply being added to the input signal.   The probability density of the noise thereby introduced is again given by Fig. 4.2a, b, or c, depending on the particular type of quantization used.

Having discussed the three basic sources of error in a digital filter, we shall now develop, as much as we can, theoretical models that make possible prediction of the amount of error.   We shall then study, in detail, several specific digital filters and compare different realizations to see which of them are less susceptible to these errors.   Some of the results will be verified experimentally, with the view of establishing confidence in the assumed models.   The main utility of these results will be a more efficient realization of the filter.   At the end of this chapter, several sections on the theory of statistical estimation of noise measurements are included; these are useful as background material for experimental verification.

4.2 ERRORS CAUSED BY INEXACT VALUES OF CONSTANT PARAMETERS

Returning first to the necessary accuracy of the constant parameters, let us consider the second-order difference equation

$$y(nT) = Ky(nT - T) - Ly(nT - 2T) + x(nT) \tag{4.3}$$

According to the analysis in Chap. 2, the z-transform $H(z)$ of the transfer function associated with (4.3) has a pole pair located at the polar coordinates r, θ in the z plane with

$$r = \sqrt{L}$$

$$\theta = \omega_r T = \cos^{-1} \frac{K}{2\sqrt{L}} \qquad (4.4)$$

We see from (4.4) that errors in the constant parameters cause a given and easily computed error in the pole positions of the digital filter. Of particular interest is the second of these equations. Parameter errors cause an error in the product $\omega_r T$. Thus, if the sampling interval T were halved, the resonant-frequency error would tend to double for comparable errors in K and L. To realize a given filter, it appears that the use of higher sampling rates requires greater accuracy of computation. This point should be carefully noted, since there seems to be a tendency to believe that higher sampling rates merely cause the digital filter to behave more like some corresponding analog filter. If a digital filter is derived, for example, by an impulse-invariant transformation from an analog filter, it is true that, *if quantization effects are ignored*, the digital filter better approximates the analog filter as T decreases. However, if the parameter error is taken into account, the opposite may happen; as T decreases beyond a certain point, the digital-filter deviation from the analog filter increases.

More explicit expression for the errors in the pole positions caused by quantization of K and L can be derived if it is assumed that the errors are small, so that [14]

$$\Delta r = \frac{\partial r}{\partial L} \Delta L + \frac{\partial r}{\partial K} \Delta K$$

$$\Delta \theta = \frac{\partial \theta}{\partial L} \Delta L + \frac{\partial \theta}{\partial K} \Delta K \qquad (4.5)$$

By using (4.4), this set of equations yields

$$\Delta r = \frac{1}{2r} \Delta L$$

$$\Delta \theta = \frac{\Delta L}{2r^2 \tan \theta} - \frac{\Delta K}{2r \sin \theta} \qquad (4.6)$$

Since $\Delta \theta = T \Delta \omega_r$, the second equation shows directly that the error sensitivity is directly proportional to the sampling rate. Furthermore, it is clear that errors in angle θ (and hence resonant frequency) are greater when θ is small so that narrowband low-pass filters are particularly sensitive to quantization effects.

The parameter-dependent nature of these quantization effects can often be lessened by using the coupled resonator described in Sec. 2.12. If we choose $A = D = r \cos \theta$ and $B = -C = r \sin \theta$, we get

$$\Delta r = \Delta A \cos \theta + \Delta B \sin \theta$$

$$\Delta \theta = -\Delta A \frac{\sin \theta}{r} + \Delta B \frac{\cos \theta}{r} \tag{4.7}$$

We see that large errors are no longer associated with small angles, as in the previous realization.

We conclude this section with an extremely important comment, namely, that it is almost never advisable to implement practical digital filters, except first- and second-order filters, in either the direct form of Fig. 2.19 or the canonic form of Fig. 2.20. For either of these two forms, the poles and zeros are extremely sensitive functions of the coefficients of the difference equation. The degree of sensitivity increases as the order of the difference equation increases. There is a tendency to assume that these effects do not become important until the order of the difference equation is high. In fact, the effect is important in third- and fourth-order systems and can even lead to instability. A fifth-order Butterworth low-pass filter with 3-db cutoff frequency at $\omega_c T = \pi/10$ will *probably* be unstable if realized with 18-bit arithmetic in the direct or canonic form [9, 10]. The remedy is to realize all digital filters as simple cascade or parallel connections of first- and second-order filters or to be extremely careful in the analysis of any higher-order system.

4.3 ERRORS CAUSED BY ANALOG–DIGITAL CONVERSION

As we showed in Sec. 4.1, errors caused by quantization† of the analog input prior to entry into the computer can be approximated by independent samples of random variables having the probability density of Fig. 4.2, with variance given by (4.2). If all other errors are ignored, the variance in the output signal $y(nT)$ may be computed by using linear-system noise theory. Since the signal and noise are independent, one can proceed with the noise computation while ignoring the signal. Let the filter be defined by the transfer function $H(z)$ and weighting function $h(nT)$. Then the output $f(nT)$, when the input consists of the noise samples $e(nT)$, can be expressed by the convolutional sum (see Sec. 2.7)

$$f(nT) = \sum_{m=0}^{n} h(mT)e(nT - mT) = \sum_{m=0}^{n} h(nT - mT)e(mT) \tag{4.8}$$

We have assumed that the input noise $e(nT)$ began at $n = 0$ and was zero before; also the output $f(nT)$ was assumed to be zero before being

† We shall assume that rounding is the type of quantization employed in this and subsequent sections.

excited by the input. Equation (4.8) can be considered the defining equation of linear discrete systems and is of particular use in dealing with linear noise theory.

To find the variance of $f(nT)$, we recall our assumption of Sec. 4.1 that each noise sample in (4.8) is uncorrelated and has variance $E_0{}^2/12$. Thus, at any time nT, the variance of $f(nT)$ is simply the sum of the variances of each term in (4.8). The variance of any given term is therefore $(E_0{}^2/12)h^2(nT - mT)$; thus the total variance is

$$\sigma_f{}^2 = \frac{E_0{}^2}{12} \sum_{m=0}^{n} h^2(nT - mT) = \frac{E_0{}^2}{12} \sum_{m=0}^{n} h^2(mT) \tag{4.9}$$

Notice that $\sigma_f{}^2$ in (4.9) is, in a sense, a time-dependent result, since it is a function of n, the number of iterations. Since $h^2(nT)$ must be positive, we see that $\sigma_f{}^2$ must increase with n from some originally minimum value. This is reasonable, since one would not expect a large variance in the output immediately after noise is applied. Physically, the variance of the output builds up and reaches an asymptote much as does a step signal applied to a linear discrete system. A steady state is always reached unless the poles of the filter lie exactly on the unit circle; that case will be treated in Chap. 5. Given that a steady state is reached, in other words, that the right side of (4.9) reaches a finite asymptote as n increases, it is possible to derive, from (4.9), another formula from which numerical results are usually more easily computed. Remembering the definition of the z-transform as given by (2.7) and recalling that $H(z)$ and $h(nT)$ are a transform pair, we can write

$$H(z)H\left(\frac{1}{z}\right) = \sum_{m=0}^{\infty} \sum_{l=0}^{\infty} h(mT)h(lT)z^{-m+l} \tag{4.10}$$

We now multiply both sides of (4.10) by $z^{-1}/2\pi j$ and perform a closed-path integration. In order to be able to interchange summation integration in (4.10), it is necessary that the path of integration be in the region of convergence of both $H(z)$ and $H(1/z)$. Following the same reasoning as used in Sec. 2.8, we can verify that the unit circle is a suitable integration path for stable filters. If the right side of (4.10) is integrated term by term, then, from the Cauchy integral theorem, all terms will be zero except those where $m = l$. The result obtained is

$$\sum_{m=0}^{\infty} h^2(mT) = \frac{1}{2\pi j} \oint H(z)H\left(\frac{1}{z}\right) z^{-1}\, dz \tag{4.11}$$

Comparing (4.9) and (4.11), we observe that the right side of (4.11) affords an alternative formula for computing the output-noise variance for the steady-state condition, i.e., only when n goes to infinity.   This expression is often easier to apply to find the variance out of specific filters because evaluation of the integral for linear discrete networks is always possible from the Cauchy residue theorem.

The reader may have observed that (4.11) had already been derived in Sec. 2.8, by using the complex convolution theorem.

4.4   ANALYSIS OF NOISE IN FIRST-ORDER FILTER CAUSED BY A–D CONVERTER

Let $h(nT) = K^n$ with $K < 1$.   It is easy to show that this $h(nT)$ is the weighting function of the digital network defined by the difference equation (4.1).   Direct evaluation of (4.9) yields

$$\sigma_f{}^2 = \frac{E_0{}^2}{12} \frac{1 - K^{2(n+1)}}{1 - K^2} \tag{4.12}$$

In the steady state, as n approaches infinity, (4.12) reduces to

$$\sigma_f{}^2 = \frac{E_0{}^2}{12(1 - K^2)} \tag{4.13}$$

To show that (4.13) is intimately related to the necessary word length of the output register which contains, successively, the values of $y(nT)$, we consider a value K equal to 0.99.   Then $\sigma_f{}^2$ is approximately $4E_0{}^2$, and the standard deviation σ_f is equal to $2E_0$.   Let us assume that for the smallest input signal that is expected to pass through the digital filter the filter output must be 40 db higher than the noise level created by quantization.   This means that the output register must have a range of at least $200E_0$.   If, furthermore, the digital filter must process data over a 40-db range of input amplitudes, this figure becomes $20{,}000E_0$.   The output-register word length required is therefore 15 bits.   Equation (4.13) has been verified experimentally, with excellent agreement [5].

It is clear from this example that quantization effects ought to be a part of the design considerations of digital filters.   In many instances of computer simulations of digital filters, the programmer may be using a 36-bit-word-length computer and can usually experimentally adjust the system gains to avoid noise effects.   In these cases, using the entire available word length of the computer carries no disadvantage.   However, in designing special digital hardware for filtering purposes, a careful specification of the required word length is a prime consideration.

A large number of digital filters contain poles very close to the unit circle.   This is certainly true for the design of highly selective filters such as those discussed in Chap. 3.   For these cases, formulas for the output-noise variance become quite simple.   Letting $K = 1 - \epsilon$ in (4.13) and assuming that ϵ^2 can be neglected, we obtain

$$\sigma_f{}^2 = \frac{E_0{}^2}{24\epsilon} \tag{4.14}$$

Thus the noise variance is inversely proportional to the distance of the pole from the unit circle and directly proportional to the gain of the filter at zero frequency.

4.5 ANALYSIS OF DIGITAL RESONATOR WITH TWO POLES AND NO ZEROS

We shall now derive formulas analogous to (4.13) for a simple digital resonator of the type discussed in Secs. 2.10 and 2.12. Consider the second-order difference equation (2.23) which has the z-transform, as indicated in (2.24),

$$H(z) = \frac{z^2}{z^2 - K_1 z - K_2} \tag{4.15}$$

The poles of $H(z)$ are located at $z = re^{\pm j\theta}$, with

$$r = \sqrt{K_2} \quad \text{and} \quad \cos\theta = \frac{K_1}{(2\sqrt{K_2})}$$

Similarly, the poles of $H(1/z)$ are located at $z = (1/r)e^{\pm j\theta}$. Thus the expression inside the integral in (4.11) can be factored, the residues found, and Cauchy's residue theorem applied to evaluate the integral. Only the result will be given here; for the $H(z)$ given by (4.15)

$$\frac{1}{2\pi j} \oint H(z)H\left(\frac{1}{z}\right) z^{-1}\, dz = \frac{1 + r^2}{1 - r^2} \frac{1}{r^4 + 1 - 2r^2 \cos 2\theta} \tag{4.16}$$

From (4.9) and from the analysis leading to the equality (4.11), we see that the expression in (4.16) is proportional to the steady-state variance of the output noise, that is, the noise obtained when n in (4.9) becomes infinite. As in Sec. 4.4, if we are discussing systems with poles close to the unit circle we can let $r = 1 - \epsilon$ and ignore terms with quadratic and higher exponents in ϵ. This leads to the formula

$$\sigma_f{}^2 = \frac{E_0{}^2}{48\epsilon \sin^2\theta} \tag{4.17}$$

Inspection of (4.17) shows that the output-noise variance is inversely proportional to the distance of the poles from the unit circle, a result that might be guessed intuitively by extension of the single-pole case. It is well known for the analog case, and equally true for the digital case, that resonator behaves like a frequency-shifted version of its low-pass proto-type when the poles are very close to the $j\omega$ axis (in the analog case) and the unit circle (in the digital case). In addition, (4.17) shows clearly the dependence of the noise on the resonant angle $\theta = \omega_r T$. For very low values of θ, the noise variance is greatly increased. A value of θ of 0.1 yields about 100 times the noise variance (10 times the standard deviation) expected if $\theta = \pi/2$. Thus the low-frequency filter produces about three or four more bits of noise standard deviation. According to the dynamic-range and signal-to-noise-ratio criteria postulated in Sec. 4.4, an 18-bit word length would be required for low-resonant-frequency filters.

4.6 ANALYSIS OF DIGITAL RESONATOR WITH TWO POLES AND ONE ZERO

Let us now consider the two-pole one-zero digital resonator defined by the difference equation (2.32) and the network shown in Fig. 2.14. The network transfer function, as indicated in (2.33), is given by

$$H(z) = \frac{z^2 - Lz}{z^2 - K_1 z - K_2} \qquad L = r\cos\theta \qquad (4.18)$$

Following the same procedure as in Sec. 4.5, we can derive the steady-state noise output variance to be

$$\sigma_f{}^2 = \frac{E_0{}^2}{12}\left[\frac{1}{1-r^2} - r^2\sin^2\theta\left(\frac{1+r^2}{1-r^2}\frac{1}{r^4+1-2r^2\cos^2\theta}\right)\right] \qquad (4.19)$$

For $r = 1 - \epsilon$, the result obtained for small ϵ is

$$\sigma_f{}^2 = \frac{E_0{}^2}{48\epsilon} \qquad (4.20)$$

Again the inverse proportionality of the noise variance to ϵ is seen, but the dependence on the resonant frequency is no longer present.

All the results obtained thus far lead to the intuitive deduction that the output-noise variance is proportional to the gain of the filter, defined at some appropriate frequency (for example, the resonant frequency of the filter). Clearly, for small ϵ, the gains of the filters thus far discussed are inversely proportional to ϵ. From very simple considerations, we should expect such behavior of the noise; after all, the higher the gain of a given system, the higher the output-noise level for a given noise input.

4.7 ERRORS CAUSED BY QUANTIZATION OF PRODUCTS

We come now to the most intricate manifestation of quantization errors, namely, errors caused by rounding off the computations used in the execution of the actual digital-filter program. Since such errors occur for each iteration of the difference equation, the effect is that of a set of noise samples superimposed on the signal; in this sense, it is similar to A-D-conversion noise. However, the precise location at which this noise is injected in the digital filter depends on the particular arrangement of the program [5, 6, 11, 12].

Figure 4.3 shows how the quantization noise caused by the multiplications is introduced into the direct realization of a four-pole four-zero system.† Associated with each multiply operation are noises $e_0(nT)$ to $e_8(nT)$. For this realization, it is clear that all noises are mutually additive and thus can be replaced by a single noise $e(nT) = \sum_{k=0}^{8} e_k(nT)$, as shown in Fig. 4.4. If we make the assumption that all noises in Fig. 4.3 are uncorrelated, then the variance of $e(nT)$ is simply nine times the variance of each small noise or $3E_0^2/4$. Note, however, that a specific program has been postulated, namely, one in which the least significant bits are thrown away after each multiplication. If the cumulative sum had been saved in a slightly larger (3 or 4 bits longer) register, the variance of $e(nT)$ could have been reduced to $E_0^2/12$. From a practical point of view, saving the extra bits could lead to an appreciably larger running time in a program, but in a special piece of hardware this worthwhile reduction could probably be accomplished at low extra cost or perhaps none at all.

An important feature of Fig. 4.3 is the passage of the noise through the filter; it is clear that this noise, unlike the A-D noise, passes only through the poles of the filter. Thus the amplification of the noise through the filter will, in general, be appreciably different from the signal amplification.

In Fig. 4.5 is shown a digital network with the same z-transform as that of Figs. 4.3 and 4.4 but in the canonic realization. We see that the noises enter the system somewhat differently from the direct form. Figure 4.6 shows an equivalent network, where

$$e_A(nT) = e_5(nT) + e_6(nT) + e_7(nT) + e_8(nT)$$

and

$$e_B(nT) = \sum_{k=0}^{4} e_k(nT)$$

It is clear that $e_A(nT)$, with variance $E_0^2/3$, passes through the entire network, both poles and zeros, whereas $e_B(nT)$ is simply a noise added to the output.

On the basis of Figs. 4.3 to 4.6, how can we compare the direct and

† As noted in Sec. 4.2, it is unwise to realize a fourth-order system in the direct form.

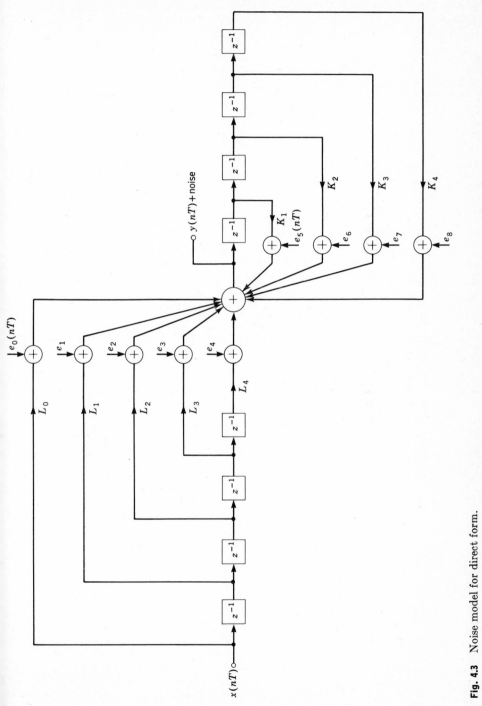

Fig. 4.3 Noise model for direct form.

110

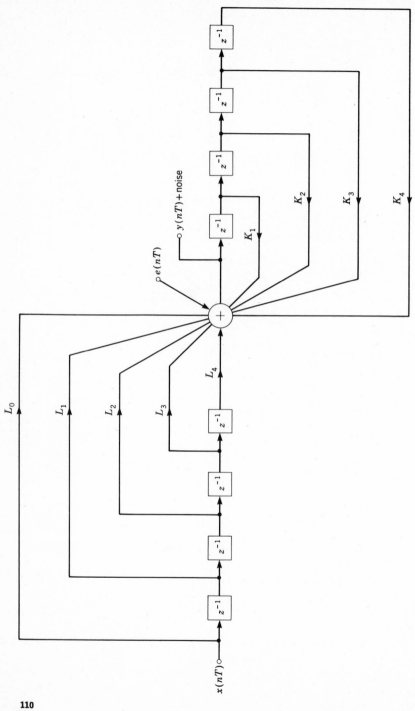

Fig. 4.4 Equivalent noise model for direct form.

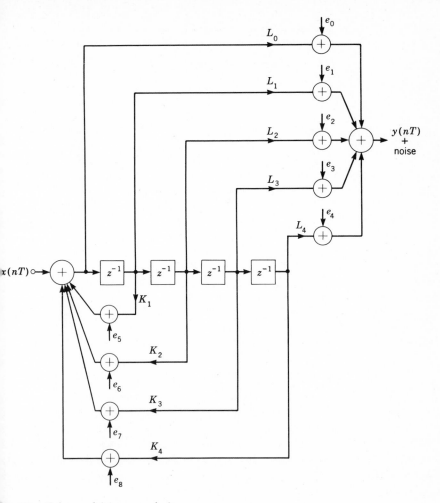

Fig. 4.5 Noise model for canonic form.

canonic realizations?   In low-pass and bandpass filters, the effect of the passband zeros is to attenuate the noise, whereas the effect of the passband poles is to accentuate it.   Therefore, it seems that the direct form introduces more than the canonic form in these cases, since the noise passes through only the high-gain poles in the former case but through both poles and low-gain zeros in the latter case.   The answer is not, however, that simple, since the low-gain zeros in the direct case mean that the signal levels through the poles will be smaller.   Since our ultimate design aim is reduced register length, we must look not only at the noise but also at signal levels.

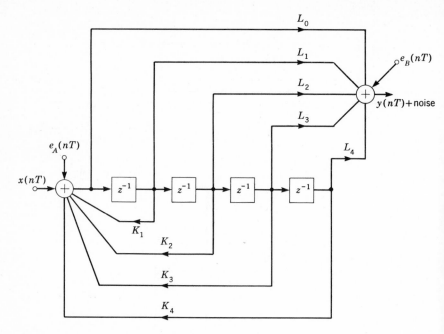

Fig. 4.6   Equivalent noise model for canonic form.

Figures 4.7 and 4.8 show cascade realizations of the fourth-order system.   Figure 4.7 is composed of two second-order direct realizations. The noises are inserted as before, to represent the multiplication errors. Figure 4.8 is a cascade of two canonic second-order filters, and the noises are shown as the combined effect of the multiply operations.   Note that, except for the beginning and end, a chain of canonic or direct cascaded resonators results in identical configurations.

4.8   DEADBAND EFFECT

In Sec. 4.7 we assumed that the errors injected due to each roundoff are mutually uncorrelated random variables.   We know of at least one counter example, namely, when the input to the filter is a constant.   Let us examine this effect in detail.

Consider the simple filter of (4.1) with $K = 0.96$ and suppose the input is a constant equal to 10.   Let the rounding be to the nearest integer ($E_0 = 1$) and suppose $y(-T) = 265$.   With precise arithmetic, we

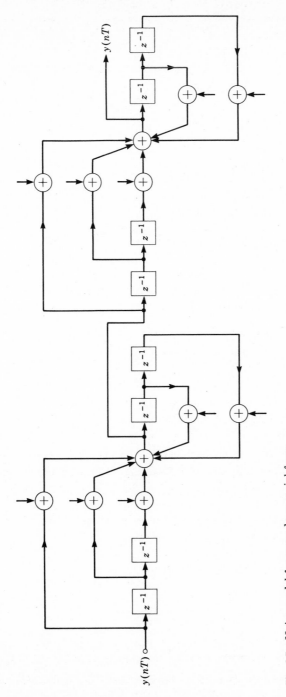

Fig. 4.7   Noise model for cascade or serial form.

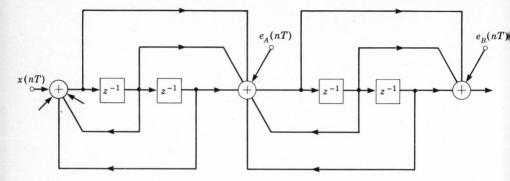

Fig. 4.8 Equivalent noise models for serial form.

expect that $y(nT)$ will approach a steady-state value of 250. With rounding, however, we observe the following:

	Before roundoff	After roundoff
$y(0)$	264.40	264
$y(T)$	263.44	263
$y(2T)$	262.48	262
$y(3T)$	261.52	262
. . . . .	. . . . . .	. . .
$y(nT)$	261.52	262

and the filter has reached a steady-state value of 262. Suppose the initial condition had been, instead, $y(-T) = 245$. Then the iterations proceed as follows:

	Before roundoff	After roundoff
$y(0)$	245.20	245
. . . . .	. . . . . .	. . .
$y(nT)$	245.20	245

and in fact any initial condition in the range $238 \leq y(-T) \leq 262$ will serve as a steady-state condition. This range is called the deadband [3].

If we consider the general case of a direct realization of a digital filter as in (2.38) and assume that all products are summed before roundoff, we can derive a general expression for the deadband effect. Let the constant input be X and suppose the output has reached a steady value Y, not necessarily equal to the predicted value Y_c, which would occur if infinite-precision arithmetic was used. Then from (2.38)

$$y(nT) = Y = \Big(\sum_{i=0}^{r} L_i \Big) X - \Big(\sum_{i=1}^{m} K_i \Big) Y + e(nT) \tag{4.21}$$

where $|e(nT)| \leq E_0/2$. This permits us to write an inequality for Y.

$$\left| \Big(1 + \sum_{i=1}^{m} K_i \Big) \left(Y - \frac{\sum\limits_{i=0}^{r} L_i}{1 + \sum\limits_{i=1}^{m} K_i} X \right) \right| \leq \frac{E_0}{2} \tag{4.22}$$

Note that the "correct" value of Y would be

$$Y_c = \frac{\sum\limits_{i=0}^{r} L_i}{1 + \sum\limits_{i=1}^{m} K_i} X \tag{4.23}$$

so that the deviation of the actual value from the "correct" value satisfies the inequality

$$|Y - Y_c| \leq \frac{E_0}{2 \left| 1 + \sum\limits_{i=1}^{m} K_i \right|} \tag{4.24}$$

Equation (4.24) says that a necessary condition for a steady-state constant error when the input is constant is that the error must satisfy the inequality. The total number of quantization steps in this deadband is approximately equal to the dc gain of the filter with poles only. The necessary condition is also a sufficient condition if the past m outputs for an mth-order filter are equal and lie within the deadband. If truncation were used instead of rounding, the width of the deadband would be the same but it would not be symmetrically distributed about Y_c.

For a first-order filter with zero input, we can conclude that the output will not decay to zero but will "hang up" as soon as the edge of the deadband is reached. Higher-order filters have more complicated effects; the output may go through the deadband and reach the other side, or steady-state oscillations may occur. There is also a phenomenon like the deadband associated with the sampling frequency. Consider the filter specified by the equation

$$y(nT) = Ky(nT - T) + Ly(nT - 2T) + x(nT)$$
$$K = 1.1 \quad L = 0.9 \tag{4.25}$$

and let the initial conditions by $y(-T) = 2$, $y(-2T) = -2$. If we quantize to the nearest integer we observe the following behavior with $x(nT) = 0$:

	Before roundoff	After roundoff	Previous output
$y(0)$	-1.6	-2	$+2$
$y(T)$	$+1.6$	$+2$	-2
$y(2T)$	-1.6	-2	$+2$
.....	.....	...	...

We observe that an oscillation is maintained at the frequency π/T. The reason is that there is a noise sample each time the equation is iterated, and the filter magnifies the frequency component at π/T until it captures a quantization level.

One can sometimes avoid the deadband effect by adding a small noise to the input of the filter [3]. This is called *dithering* and is sometimes very effective.

4.9 ROUNDOFF-NOISE FORMULAS FOR DIFFERENT ARRANGEMENTS OF DIGITAL NETWORKS

With the assumption of uncorrelated noise samples, the general theory of Sec. 4.3 developed for A-D-conversion noise can be directly applied to the roundoff problem. Indeed, the only difference between roundoff and A-D noise is topological; the A-D noise enters at the filter input whereas roundoff can enter at other points in the system. The basic formulas (4.9) and (4.11) can be directly applied, provided that one uses the transfer function that correctly relates the noise sources to the resultant output noise created by that source. For example, for Fig. 4.3, let us write the total function as

$$H(z) = \frac{R(z)}{P(z)} \tag{4.26}$$

Clearly $R(z)$ defines the zeros of $H(z)$ created by the four feedforward delays and $P(z)$ defines the poles of $H(z)$ created by the four feedback delays. Inspection of Figs. 4.3 and 4.4 shows that the noise cannot be affected by $R(z)$. Thus the transfer function relating the noise source $e(nT)$ of Fig. 4.4 and the subsequent steady-state output noise is given by

$$\sigma_1{}^2 = \frac{mE_0{}^2}{12(2\pi j)} \oint \frac{z^{-1}}{P(z)P(1/z)} \, dz \tag{4.27}$$

where we have assumed that the variance of the noise source created by each noise generator is $E_0{}^2/12$ and that there are m such noise generators.

In Fig. 4.6, the noise $e_A(nT)$ passes through both the poles and the zeros, that is, through the entire filter, so that the variance of the output noise due to $e_A(nT)$ becomes

$$\sigma_2{}^2 = \frac{mE_0{}^2}{12(2\pi j)} \oint H(z)H\left(\frac{1}{z}\right) z^{-1}\, dz \tag{4.28}$$

In this case, the roundoff noise passes through the same filter as the A-D noise and is thus *additive* to the input signal. The total noise is obtained by adding the variance of $e_B(nT)$ to $\sigma_2{}^2$.

In the cascade form shown in Fig. 4.7, let us write $H(z)$ in the form

$$H(z) = \frac{R_1(z)}{P_1(z)} \frac{R_2(z)}{P_2(z)} \tag{4.29}$$

where $R_1(z)$ and $P_1(z)$ refer to the zeros and poles of the first resonator and $R_2(z)$ and $P_2(z)$ to the zeros and poles of the second resonator. If each noise has variance $E_0{}^2/12$, the resultant output variance is

$$\sigma_3{}^2 = \frac{5E_0{}^2}{12(2\pi j)} \oint \left[\frac{R_2(z)R_2(1/z)}{P_1(z)P_1(1/z)P_2(z)P_2(1/z)} \right.$$
$$\left. + \frac{1}{P_2(z)P_2(1/z)} \right] z^{-1}\, dz \tag{4.30}$$

In the form shown in Fig. 4.8, the result is

$$\sigma_4{}^2 = \frac{E_0{}^2}{12(2\pi j)} \left\{ \oint \left[\frac{2R_1(z)R_1(1/z)R_2(z)R_2(1/z)}{P_1(z)P_1(1/z)P_2(z)P_2(1/z)} \right. \right.$$
$$\left. \left. + \frac{5R_2(z)R_2(1/z)+2}{P_2(z)P_2(1/z)} \right] z^{-1}\, dz \right\} \tag{4.31}$$

4.10 EXAMPLE: DIFFERENT ARRANGEMENT OF TWO-POLE ONE-ZERO NETWORK

Equations (4.28) to (4.31) at least imply that the effects of roundoff noise vary with the specific configuration chosen, even though all the configurations have the same overall transfer function. We now seek some insight into the problem of choosing the configuration that minimizes the bad effects of this noise. Thus, given an $H(z)$, one needs to evaluate equations of the type (4.28) to (4.31) but to do so is usually a laborious undertaking. However, some practice with digital networks should make possible good judgment as to the most desirable network arrangement. Let us consider a rather simple example, which illustrates some of the basic considerations.

The transfer function

$$H(z) = \frac{A_1 z}{z - \beta} + \frac{A_2 z}{z - \gamma} \tag{4.32}$$

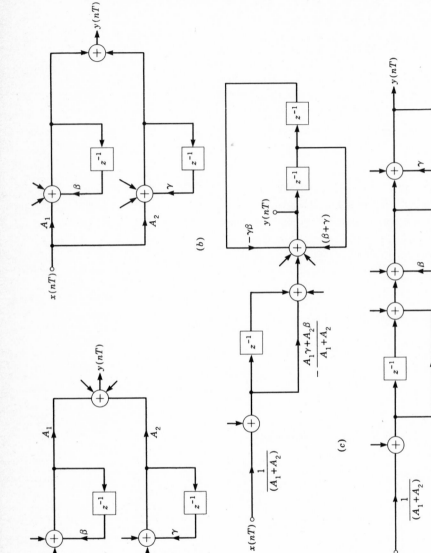

Fig. 4.9 Noise models for various forms of same network.

where β and γ are real poles of the system, of magnitude less than unity, can be realized in a variety of ways, some of which are shown in Fig. 4.9.

We see that Fig. 4.9a and b differ in a seemingly trivial way, namely, the multiplying factors A_1 and A_2 appear either before or after the two parallel first-order iterations. In these figures, each unlabeled arrow indicates the insertion of a noise source of variance $E_0{}^2/12$. The steady-state output noise is easily computed by using (4.13); for Fig. 4.9a,

$$\sigma_1{}^2 = \frac{E_0{}^2}{12} \left(\frac{A_1{}^2}{1 - \beta^2} + \frac{A_2{}^2}{1 - \gamma^2} + 2 \right) \tag{4.33}$$

and for Fig. 4.9b,

$$\sigma_2{}^2 = \frac{E_0{}^2}{6} \left(\frac{1}{1 - \beta^2} + \frac{1}{1 - \gamma^2} \right) \tag{4.34}$$

Thus the desirability of placing the multiplication by A_1 and A_2 before or after the iterations depends entirely on the values of A_1 and A_2; if they are much greater than unity, Fig. 4.9a yields less noise; if much less than unity, Fig. 4.9b yields less noise. Notice that the A-D noise does not depend on the placement of A_1 and A_2; thus (4.33) and (4.34) tell us all we need to know about the relative noises.

The arrangements of Fig. 4.9c and d are somewhat less trivial to analyze. Considering first Fig. 4.9c, we can see that three of the four noises pass through the transfer function

$$F(z) = \frac{z^2}{(z - \gamma)(z - \beta)} \tag{4.35}$$

Thus,

$$\sigma_3{}^2 = \frac{E_0{}^2}{12(2\pi j)} \oint F(z) F\left(\frac{1}{z}\right) z^{-1} \, dz \tag{4.36}$$

which can be evaluated to yield

$$\sigma_3{}^2 = \frac{E_0{}^2}{12} \left[\frac{1 + \beta\gamma}{1 - \beta\gamma} \frac{1}{(1 - \beta^2)(1 - \gamma^2)} \right] \tag{4.37}$$

where $\sigma_3{}^2$ is the output-noise variance caused by three of the four noise sources shown in Fig. 4.9c and is therefore less than the total output noise of that network.

It is instructive to compare (4.34) and (4.37). The ratio of the variances is

$$\frac{\sigma_2{}^2}{\sigma_3{}^2} = \frac{(2 - \gamma^2 - \beta^2)(1 - \beta\gamma)}{1 + \beta\gamma} \tag{4.38}$$

If, as is true in many practical cases, the poles β and γ are both close to the unit circle, it is clear from (4.38) that $\sigma_2{}^2$ is smaller than $\sigma_3{}^2$. Thus

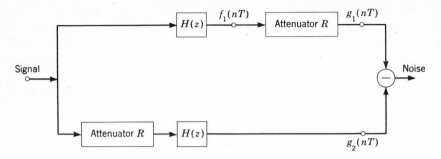

Fig. 4.10   Experimental setup for noise measurement.

the parallel form of Fig. 4.9*b* is superior to the direct form of Fig. 4.9*c*. Whether or not this situation remains if the multiplier $1/(A_1 + A_2)$ and the zero of Fig. 4.9*c* are moved to the right of the second-order iteration depends on the numerical values of A_1, A_2, γ, and β.

4.11 NOISE MEASUREMENTS

Thus far in this chapter we have dealt with a specific problem of significance for the practical design of digital filters and have tried to avoid too much mathematical terminology in order to focus more strongly on the physical phenomena. However, quantization problems encompass a large area of important issues, most of which have not yet been resolved. It thus becomes important to devise experimental ways of studying such digitally caused noise problems, and, for this, some mathematical background is required. This background consists primarily of the theory of spectral estimation for the discrete case. Much literature [2, 4, 7, 8] is available on this subject; the present section summarizes some of this information.

One way to measure the noise generated via quantization in a digital filter is illustrated in Fig. 4.10. Two identical filters $H(z)$ are used, but the top $H(z)$ is excited by a large enough signal so that $f_1(nT)$ has a large signal-to-noise ratio such that, after passing through the attenuator R, it is reasonable to assume that the amount of noise embedded in $g_1(nT)$ is very small. Thus, if the noise generated by $H(z)$ is significant, nearly all the noise is to be found in $g_2(nT)$ and therefore the difference measurement is a direct measure of the noise created by $H(z)$. In general, one wants to measure the statistical properties of this noise, such as its probability distribution, mean value, root-mean-square value, correlation function, and spectrum; since all these are statistical measurements, it is necessary to understand the functional dependence of the measurement errors on the specific measurement used.

4.12 ESTIMATION OF THE MEAN VALUE OF THE NOISE

The concepts of probability distribution, stationary random process, the ergodic hypothesis, the central limit theorem, and gaussian processes are essentially the same whether applied to continuous or discrete processes, and we assume that the reader has a reasonable grasp of these topics. Let us then inquire into a few simple statistical problems, developing our definitions as we go.

We consider first the problem of estimating the ensemble mean value of a finite piece of noise data containing $N + 1$ samples of data. The natural way to define this estimate is

$$A[x(nT)] = \frac{1}{N + 1} \sum_{n=0}^{N} x(nT) \tag{4.39}$$

The random variable A is a good estimate if its ensemble mean value is equal to the ensemble mean value of the original noise $x(nT)$ and if its variance becomes zero for large enough N. The first condition is obviously satisfied provided $x(nT)$ is stationary so that, for example, its ensemble mean is time-invariant. To study the second condition, let us write $\sigma^2[A] = E[A^2] - E^2[A]$ for the variance, where E represents the ensemble mean value. Then

$$\sigma^2[A] = \frac{1}{(N + 1)^2} \sum_{n=0}^{N} \sum_{m=0}^{N} \{E[x(nT)x(mT)] - E^2[x(nT)]\} \tag{4.40}$$

Again, if $x(nT)$ is stationary, the term inside the double summation is a function of only the time difference $nT - mT$ and can be written as $Q(nT - mT) = Q(mT - nT)$. This leads via (4.40) to the equation

$$\sigma^2[A] = \frac{1}{N + 1} \sum_{n=-N}^{N} \left(1 - \frac{|n|}{N + 1}\right) Q(nT) \tag{4.41}$$

Equation (4.41) establishes a constraint on the *autocovariance* $Q(nT)$ of the original process, so that if $\lim_{N \to \infty} \sigma^2[A] = 0$ then A can be said to be a good estimate. As an example, let $Q(nT) = K^{|n|}$, where $K < 1$ so that $Q(nT)$ decreases with increasing n. Then (4.41) can be evaluated to give

$$\sigma^2[A] = \frac{1}{N + 1} \left\{ \frac{1 + K - 2K^{N+1}}{1 - K} - \frac{1}{N + 1}\left[\frac{2K(1 - K^N)}{(1 - K)^2} - \frac{2NK^{N+1}}{1 - K}\right]\right\} \tag{4.42}$$

For large N, terms of the form K^N can be ignored, so that (4.42) reduces to

$$\sigma^2[A] \approx \frac{1 + K}{(N + 1)(1 - K)} \tag{4.43}$$

From (4.43), it is easy to estimate how large N must be for a particular K so that $\sigma^2[A]$ is satisfactorily small.

Statistical parameters often may be measured via linear filtering. Thus, for example, to estimate the mean value of a signal, rather than performing the measurement (4.39), the noise $x(nT)$ may be passed through the first-order difference equation,

$$y(nT) = y(nT - T) + K[x(nT) - y(nT - T)]$$

If the noise, having a fixed mean value m_0, suddenly excites this first-order system beginning at $n = 0$, it is easy to see that the mean value of the output $y(nT)$ can be expressed as

$$E[y(nT)] = m_0 \sum_{k=0}^{n} h(kT) = m_0[1 - (1 - K)^{n+1}]$$

Thus the mean value of the filter output asymptotically approaches the mean value of the input. Now, it needs to be proved that the variance of the output asymptotically approaches zero. This is indeed true if the covariance function $Q(nT)$ of (4.42) is used, namely, $Q(nT) = K^{|n|}$. The proof is left to the reader.

Not only the ensemble mean value but also the spectral density of a random function may be estimated by using filters. This will be seen more clearly in the next section.

4.13 THE AUTOCOVARIANCE AND SPECTRAL DENSITY

The autocovariance $Q(nT)$ is a two-sided sequence, having values for both positive and negative n and also having the special property that $Q(nT)$ is an even function of nT. The spectral density can thus be defined as the two-sided z-transform

$$S(z) = \sum_{n=-\infty}^{\infty} Q(nT)z^{-n} \tag{4.44}$$

The functions $Q(nT)$ and $S(z)$ are the pertinent transform pair for the analysis of noise, corresponding to the wave $x(nT)$ and its transform $X(z)$ for determinate waveforms. To better appreciate these entities, an examination of the effect of a linear filter on the noise will be useful. By using (4.44), the autocovariance of the output $y(nT)$ of a linear filter can be expressed in terms of its input $x(nT)$, as follows:

$$E[y(nT)y(mT)] = \sum_{k=0}^{n} \sum_{l=0}^{m} Q_x(nT - mT - kT + lT)h(kT)h(lT)$$

$$\tag{4.45}$$

where

$$Q_x(kT - lT) = E[x(kT)x(lT)]$$

In (4.45) it has been assumed that the process has been applied suddenly to the filter at $n = 0$; thus $y(nT)$ does not become a stationary process until the natural oscillations of the filter have died. When this steady state has been reached, it is possible to write

$$Q_y(nT - mT) = E[y(nT)y(mT)]$$
$$= \sum_{k=0}^{\infty} \sum_{l=0}^{\infty} Q_x(nT - mT - kT + lT)h(kT)h(lT) \quad (4.46)$$

Equation (4.46) can be formulated in spectral terms by multiplying both sides by z^{-n+m} and summing over doubly infinite limits. The triple sum now becomes separable, resulting in the well-known formula

$$S_y(z) = S_x(z)H(z)H\left(\frac{1}{z}\right) \quad (4.47)$$

which is, of course, analogous to (2.41). Notice, however, that (4.47) involves two-sided z-transforms, that it is a power relationship, and that the effective *power transfer function* $H(z)H(1/z)$ has poles both inside and outside the unit circle.

From the definition of $Q(nT)$, it is apparent that $Q(0)$ is the mean-squared value of the random process, namely,

$$Q(0) = E[x^2(nT)] \quad (4.48)$$

But by the inversion theorem,

$$Q(kT) = \frac{1}{2\pi j} \oint S(z)z^{k-1}\, dz \quad (4.49)$$

If we let $z = e^{j\omega T}$ and $\omega = 2\pi f$, (4.49) becomes, for $k = 0$,

$$Q(0) = T \int_0^{2\pi} S(e^{j2\pi fT})\, df \quad (4.50)$$

Equation (4.50) shows that the mean-squared value of a random process is the integrated value of the spectral density once around the unit circle, thus justifying use of the term "spectral" density for $S(z)$.

4.14 ESTIMATION OF THE COVARIANCE FUNCTION AND SPECTRAL DENSITY

The covariance function $Q(kT)$ can be estimated by the same general procedure as that of estimating the mean value, namely, measuring the time average. Thus, if we have $N + 1$ data samples of the original process $x(nT)$ and we want an estimate of $Q(kT)$, we consider the product

$x(nT)x(nT + kT)$ and find the time average of this product. If it is assumed that $x(nT)$ is zero outside the measurement interval, the measurement is

$$R_N(kT) = \frac{1}{N - |k| + 1} \sum_{n=0}^{N-|k|} x(nT)x(nT + |k|T) \qquad (4.51)$$

As before, one can show that the mean value of the measurement $R_N(kT)$ is exactly the covariance and, also, one can find the conditions that will cause the variance of $R_N(kT)$ to become zero as $N \rightarrow \infty$; in fact, we can see from (4.51) that this condition depends on the fourth moment of the original random process.

Thus far, estimation of statistical parameters appears straightforward, and, in fact, both the mean and covariance estimates could have been deemed consistent by appealing to the ergodic hypothesis. Estimation of the spectral density may, however, lead one into error if special care is not taken. Consider, for example, what appears to be a natural estimate of $S(z)$. Let

$$X_N(z) = \sum_{n=0}^{N} x(nT)z^{-n} \qquad X_N\left(\frac{1}{z}\right) = \sum_{n=0}^{N} x(nT)z^n \qquad (4.52)$$

and define the spectral estimate

$$S_N(z) = \frac{1}{N + 1} X_N(z)X_N\left(\frac{1}{z}\right) \qquad (4.53)$$

$S_N(z)$ is commonly known as the *periodogram*.

We can first show, by manipulation of (4.52) and (4.53) and using the definition (4.51), that

$$S_N(z) = \sum_{m=-N}^{N} \left(1 - \frac{|m|}{N + 1}\right) R_N(mT)z^{-m} \qquad (4.54)$$

Now a strange thing happens. If, in (4.54), we take the limit as $N \rightarrow \infty$, then the right side of (4.54) reduces to the right side of (4.44) (if ergodicity holds), so that $S_N(z) \rightarrow S(z)$. However, if in (4.53) we first compute the variance of $S_N(z)$, then, as we let $N \rightarrow \infty$, this variance does not approach zero. A simple example will suffice to demonstrate this point. Let us consider the spectral estimate of (4.53) for $z = 1$, that is, zero frequency. Then,

$$E[S_N{}^2(1)] = \frac{1}{(N + 1)^2} \sum\sum\sum\sum E[x(nT)x(mT)x(lT)x(kT)] \qquad (4.55)$$

If we consider a gaussian process, the fourth moment is determined from the second moments by

$$E[x_1x_2x_3x_4] = E[x_1x_2]E[x_3x_4] + E[x_1x_3]E[x_2x_4] + E[x_1x_4]E[x_2x_3] \qquad (4.56)$$

From (4.55) and (4.56), it can be shown that the variance of $S_N(1)$ is $\sqrt{2}$ times the mean of $S_N(1)$ for all N; thus the variance does not approach zero, and the estimate is not consistent.

The physical interpretation of the above arguments is as follows: In making a *direct* spectral measurement via (4.53), the variance in the measurement can be reduced by performing the same measurement over different portions of the data and then averaging. As an example, say we divide N samples into M equal parts, each containing N/M samples. If N/M is kept fixed while both increase, it is clear from the ergodic hypothesis that convergence in the mean results. This measurement can be formulated as follows: Let

$$X_N^k(z) = \sum_{kN}^{(k+1)N-1} x(nT)z^{-n} \tag{4.57}$$

$$S_N^k(z) = \frac{X_N^k(z)X_N^k(1/z)}{N} \tag{4.58}$$

$$F_{N,K}(z) = \frac{1}{K} \sum_{k=0}^{K-1} S_N^k(z) \tag{4.59}$$

By the ergodic hypothesis,

$$\lim_{N\to\infty} \lim_{K\to\infty} F_{N,K}(z) \to S(z) \tag{4.60}$$

Rather than using a direct spectral measurement, we can resort to (4.54), first measuring the correlation function $R_N(mT)$, using (4.51). Since (4.54) and (4.53) are equivalent measurements, it is clear that direct use of (4.54) over all samples will result in a large measurement error. Physically, we can see that, as $k \to N$ in (4.51), the estimate $R_N(kT)$ becomes poor, since so few terms are used in the estimate; in fact, $R_N(NT)$ is based on the single product $x(0)x(NT)$ and this means that the variance of that measurement may be N times the variance of $R_N(0)$ (this would be the case if the original random processes consisted of uncorrelated variables). From this we infer that a spectral measurement based on (4.54) should derive primarily from the values of $R_N(mT)$ for m small compared with N; this can be done in general by incorporating a *window* $W(mT)$ in (4.54),

$$T_N(z) = \sum_{m=-N}^{N} \left(1 - \frac{|m|}{N+1}\right) W(mT)R_N(mT)z^{-m} \tag{4.61}$$

In general we would want $W(mT)$ to be an even function of m and to decrease in value with increasing $|m|$ so that, by the time $m \approx N$, $W(mT)$ is negligible. A particular case of a window is $W(mT) = 0$ for

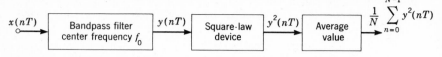

Fig. 4.11   Measurement of average power in frequency band.

$|m| \geq M$ and $W(mT) = 1$ for $|m| < M$. Physically, we see that, as M is decreased, the spectral resolution of the measurement must decrease and the measurement itself increases in accuracy.

By means of the complex convolution theorem (2.15), it is possible to associate a weighting of the spectral measurement with $W(mT)$. Thus, if $U(z)$ is the two-sided z-transform of $W(mT)$, then

$$T_N(z) = \frac{1}{2\pi j} \oint S_N(v) U\left(\frac{z}{v}\right) v^{-1} \, dv \tag{4.62}$$

$T_N(z)$ is seen to be the weighted average (over frequency) of the spectral density $S_N(v)$. If $W(mT)$ is "wide" (extends over many samples), $U(z)$ is "narrow" so that the integral (4.62) is approximately the spectral density $S_N(z)$, giving good frequency resolution but, as we have seen, large statistical fluctuations.

In addition to the direct and indirect (via the correlation function) methods of spectral estimation, there is a third technique, using a filter bank, as shown in Fig. 4.11.

If a set of n of these channels is used, then n points on the spectral-density curve can be estimated. A simple intuitive picture of the action of the configuration of Fig. 4.11 is obtained by supposing that the signal is white noise, that the bandpass filter is a narrow filter with ideal, square, cutoff properties and bandwidth B, and that the spectrum is constant over the bandwidth of this filter. Then from (4.50)

$$Q_y(0) = E[y^2(nT)] = BTS(e^{j\omega_0 T}) \tag{4.63}$$

But, by the ergodic hypothesis,

$$\lim_{N \to \infty} \frac{1}{N} \sum_{n=0}^{N-1} y^2(nT) = E[y^2(nT)]$$

so that the measurement is a consistent estimate of the spectrum at $\omega_0 = 2\pi f_0$.

We have thus defined three ways of performing a spectrum analysis, but we have not treated the relationships between the values of the measurement parameters and the resultant variances of the measurement, nor have we discussed the probability distribution of these spectral measurements. The reader interested in further studying these questions for discrete variables is referred to the somewhat formidable works of Grenander and Rosenblatt [7] and of Hannan [8].

4.15 EXAMPLE: MEASUREMENT OF NOISE IN A RESONATOR CASCADE

The simple model of quantization noise which was introduced in Sec. 4.3 and which led to the general result (4.9) is based on the idea that the noise samples generated are uncorrelated.   For such a hypothesis experimental verification was deemed desirable.   Therefore, a particular configuration, shown in Fig. 4.12, was studied [6].

The device of Fig. 4.12 is actually a vowel generator commonly used in speech research; if the resonant frequency of $H_1(z)$ is set to 200 Hz and the resonant frequency of $H_5(z)$ is set to 3,500 Hz and that of $H_6(z)$ to 4,500 Hz, then the 10 frequency settings for H_2, H_3, and H_4 to yield 10 American vowels are shown in Table 4.1, the vowel being produced by the application of a periodic pulse train to the input of the network.   The period of the pulse train is within the range of the expected period between successive vocal-cord epochs in human speech production; in our measurements, a period of 8 msec was used.

The mean-squared noise created by this system for each of the 10 vocal settings was measured in the manner indicated in Fig. 4.10.   Since the input pulse train was not derived from an analog signal but was simply internally generated, no A-D noise but only roundoff noise was

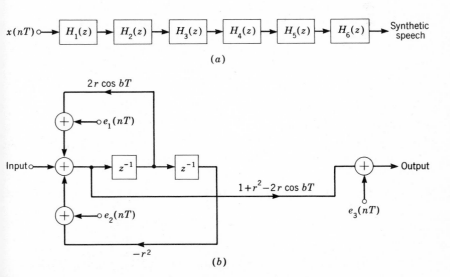

(a)

(b)

Fig. 4.12   Digital formant network. (a) Cascade of six resonators; (b) configuration of each $H(z)$.

Table 4.1   Resonator settings for 10 vowels

IPA symbol	Typical word	H_2	H_3	H_4
i	(beet)	270	2,290	3,010
I	(bit)	390	1,990	2,550
ε	(bet)	530	1,840	2,480
ae	(bat)	660	1,720	2,410
Λ	(but)	520	1,190	2,390
a	(hot)	730	1,090	2,440
$\circ$	(bought)	570	840	2,410
U	(foot)	440	1,020	2,240
μ	(boot)	300	870	2,240
3	(bird)	490	1,350	1,690

generated by this system.   According to our hypothesis, it is permissible to model the noise by assuming independent noise sources as shown in Fig. 4.6b and adding the noise powers contributed by each of the sources. Thus, for example, if the network of Fig. 4.12b were $H_2(z)$ then $e_1(nT)$ passes through $H_2H_3H_4H_5H_6 = H_{26}(z)$; the noise created this way is, from (4.9),

$$\sigma_2{}^2 = \frac{E_0{}^2}{12} \sum_{m=0}^{\infty} h_{26}{}^2(mT) \tag{4.64}$$

where $h_{26}(mT)$ is the inverse transform of $H_{26}(z)$.   Similarly, $\sigma_3{}^2$ can be found via the impulse response $h_{36}(mT)$; written compactly, the total noise caused by single noise sources in each $H(z)$ is

$$\sigma_0{}^2 = \sum_{i=2}^{6} \sigma_i{}^2 = \frac{E_0{}^2}{12} \sum_{i=2}^{6} \sum_{m=0}^{\infty} h_{i6}{}^2(mT) \tag{4.65}$$

Since, as seen in Fig. 4.12b, there are three noise sources per $H(z)$, the theoretical noise power would be $3\sigma_0{}^2$.   Table 4.2 compares the measurement of $3\sigma_0{}^2$ as obtained from (4.5) and the measurement made via Fig. 4.10, each of these measurements being based on 3,500 samples, for the resonator settings of Table 4.1. E_0 is taken as unity.   It is seen that the agreement is not bad and certainly justifies the use of our model to establish criteria for register length.

4.16   SUMMARY

This chapter has considered the three sources of error when a recursive digital filter is designed on the basis of perfect accuracy and implemented with finite-accuracy arithmetic.   One effect is quantization of the input,

Table 4.2 Measured vs. predicted noise for 10 vowels

IPA symbol	Typical word	Measured noise	Theoretically determined value
i	(beet)	702,664	735,547
I	(bit)	125,574	114,717
ε	(bet)	101,110	104,036
ae	(bat)	57,674	52,241
Λ	(but)	51,414	52,241
a	(hot)	51,050	53,346
$\mathfrak{o}$	(bought)	50,700	52,763
U	(foot)	53,460	52,242
μ	(boot)	41,044	42,123
3	(bird)	27,574	27,465

which was treated as additive random noise.   We derived an expression for the output of a digital filter with noise as input.   A second effect is caused by the rounding or truncation of products or sums of products that are to be fed back for use in further computations.   These errors are filtered by those further computations, and the filtering action can amplify the roundoff error considerably, depending on the form of the realization. We discussed two common situations: (1) when the errors were completely uncorrelated from sample to sample, in which case the analysis of a filter with noise input enabled us to predict the mean-squared output noise or the power spectrum of such noise, and (2) when the errors were completely correlated from sample to sample, in which case we observed a deadband or steady-state error of many times a quantization step.   The other source of error considered was the change in pole locations due to the quantization of the coefficients of the different equations.   Again the form of the realization of the filter is an important consideration.   On an average, the direct and canonic forms are highly sensitive to this effect whereas the parallel and serial forms are less sensitive.

We then saw that the estimation of noise power spectra was an area in which great care was necessary, as at least one of the intuitively sensible estimates tends to have a variance that does not become small as the measurement is refined.   In making such spectral measurements one must always be willing to measure the noise power in a band of frequencies rather than at a single frequency—the wider bandwidth is a price paid for reduced variance of the estimate.

REFERENCES

1. Bennett, W. R.: Spectra of Quantized Signals, *Bell System Tech. J.*, **27**: 446–472 (July, 1948).

2. Blackman, R. B., and J. W. Tukey: "The Measurement of Power Spectra, sec. B.5, pp. 95–100, Dover Publications, Inc., New York, 1958.

3. Blackman, R. B.: "Linear Data-smoothing and Prediction in Theory and Practice," Addison-Wesley Publishing Company, Inc., Reading, Mass., 1965.

4. Cooley, J. W., P. Lewis, and P. Welch: The Fast Fourier Transform and It Applications, *IBM Res. Paper RC*-1743, Feb. 9, 1967.

5. Gold, B., and C. M. Rader: Effects of Quantization Noise in Digital Filter, presented at 1966 Spring Joint Computer Conference, *AFIPS Proc.*, 28 213–219 (1966).

6. Gold, B., and L. R. Rabiner: Analysis of Digital and Analog Formant Synthe sizers, presented at the 1967 Conference on Speech Communications, *IEEE Trans. Audio*, **AU16**(1): (March, 1968).

7. Grenander, U., and M. Rosenblatt: "Statistical Analysis of Stationary Tim Series," John Wiley & Sons, Inc., New York, 1967.

8. Hannan, E. J.: "Time Series Analysis," John Wiley & Sons, Inc., New York, 196(

9. Kaiser, J. F.: Some Practical Considerations in the Realization of Linear Digita Filters, *Proc. 3rd Allerton Conf.*, Oct. 20–22, 1965.

10. Kaiser, J. F.: Digital Filters, in F. F. Kuo and J. F. Kaiser (eds.), "Systen Analysis by Digital Computer," chap. 7, John Wiley & Sons, Inc., New York 1966.

11. Knowles, J. B., and R. Edwards: Effects of a Finite Word Length Computer in Sampled-data Feedback System, *Proc. Inst. Elec. Engrs. (London)*, **112**(6) (June, 1965).

12. Knowles, J. B., and R. Edwards: Complex Cascade Programming and Associate Computational Errors, *Electronics Letters*, **1**(6): 160–161 (August, 1965).

13. Knowles, J. B., and E. M. Olcayto: Coefficient Accuracy and Digital Filte Response, *IEEE Trans. Circuit Theory*, **15**(1): 31–41 (March, 1968).

14. Rader, C. M., and B. Gold: Effects of Parameter Quantization on the Poles of Digital Filter, *Proc. IEEE*, **55**(5): 688–689 (May, 1967).

15. Sandberg, I. W.: Floating-point-roundoff Accumulation in Digital-filter Realiza tions, *Bell System Tech. J.*, **46**: 1775–1791 (October, 1967).

5

Block-diagram Representation
of Computer Programs

5.1 INTRODUCTION

The three preceding chapters have considered in great detail the design
and analysis of programs that execute linear difference equations. These
programs have been shown to behave like linear filters. The inputs and
outputs are thought of as samples of a time waveform, and time is
"updated" each time the difference equations are executed once. A
major application of such difference equations is in the digital-computer
simulation of analog or digital waveform-processing systems. Digital
simulation of systems is, of course, a field far too vast to be covered
adequately in a single chapter; we shall be concerned here only with a
small and easily circumscribed topic, a technique for simulating systems
by computing the values of all the waveforms of interest in a system, at
equally spaced instants of time, using only the values of these waveforms
and of a few auxiliary quantities known from the previous sampling
instant. These waveform values and auxiliary quantities can be said to
characterize the "state" of the system, and we can therefore call the

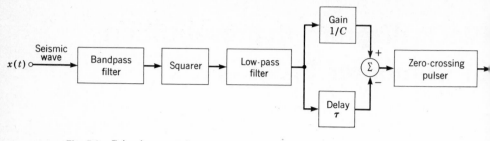

Fig. 5.1 Seismic-event detector.

simulation technique *next-state* simulation.   The computer program that
simulates a given system by next-state techniques bears a simple relationship to the system itself, although not necessarily an equivalence relationship.   The connection between the two is made most evident by the use
of a block-diagram description of the program, which is almost identical
to the block diagram familiar to the systems engineer.   This block diagram
is the easiest and most effective way to describe a next-state simulation
program, and for such a program it supersedes the flow-chart representation more familiar to computer programmers.

  We are considering here simulation of waveform-processing systems.
Before defining terms or making generalizations we should consider a
case in point.   In Fig. 5.1 there is a description of an event detector [1]
which might be used to signal the beginning of a seismic "event," e.g.,
an earthquake.   The algorithm proposed is to observe the time-varying
energy of a band-limited seismic signal, as measured by a squarer followed
by a low-pass filter, and to declare an event when that energy becomes
greater than C times the energy τ seconds earlier.   Since there is always
some seismic noise, and since events are relatively rare, the device of Fig.
5.1 is of considerable interest to seismologists.   The purpose of a simulation would be to test the workability of the whole idea, to optimize
the parameters τ and C, and to determine the parameters of the low-pass
and bandpass filters.

  The first step in the simulation process would be to choose an appropriate sampling rate, $1/T$, and to get some samples of the input into a
computer.   These samples could then be fed to a simulation program one
sample at a time.   The simulation program would take the input sample
$x(nT)$, compute the bandpass-filter output $y(nT)$, compute $y^2(nT)$, and
use it as the input to a digital low-pass filter.   The output of the low-pass
filter would then be used as the input to a simulated delay element and a
simulated gain circuit.   The output of the delay line would then be
subtracted from the output of the gain circuit, and a pulse produced if the
difference was positive where it had previously been negative, thus indi-

cating the presumed detection of an event.   If no such success occurred on the nth sample, the next sample, $x(nT + T)$, would be requested, and the whole process would be repeated, then again for $x(nT + 2T)$, etc. If all the calculations required for one sample could be completed in a time less than T, it would be possible for the program to process data continuously; otherwise a reasonably large amount of data would be saved in the memory and processed.

In what follows, we shall distinguish between the system that the engineer wishes to simulate and the approximation to the system that the computer simulation actually achieves.   The system in the engineer's mind will be called the *source system*, and the system actually programmed will be called the *object system*.   Although the object system and the source system are not generally equivalent—for example, the source system might not be discrete whereas the object system must be discrete— they should correspond fairly closely for the simulation to be useful. For example, where the source system has an analog filter, the object system will have a digital filter.   The terms source system and object system are chosen to be reminiscent of the terms *source program* and *object program* used to describe compilers.   Later in this chapter we shall consider compilers that accept as source program a description of a source system and produce an object program equivalent to an object system corresponding to that source system.   We make a fine distinction here between an object system corresponding to a source system and an object program equivalent to an object system.

For the example of digital filters and difference equations, a program to execute difference equations is *equivalent* to a digital filter such as one could construct with digital components.   The digital filter, however, can only *correspond* to some analog filter with nearly the same properties. To decide how much of a correspondence between source and object systems is needed in a given simulation project is a task requiring judgment, and the engineer cannot be relieved of this task.

5.2 THE OBJECT SYSTEM

In this section we discuss the characteristics of computer programs that follow the rules of the next-state technique.   We shall avoid any questions of validity, that is, of correspondence between source and object systems; rather, we are concerned with the connection between the object system, the computer program, and a block-diagram representation. By way of illustration, we consider a program that might be written to measure the roundoff noise generated by a first-order digital filter.   The program measures the mean-squared difference between the outputs of two digital filters identical except for roundoff.   One filter uses arithmetic much more precise than the other and serves as a reference, considered noise-free. A flow chart for such a program is shown in Fig. 5.2a. Box B generates a sample of the common input to the two filters, box C executes an iteration of the difference equation representing the reference filter, and box D executes an iteration of the difference equation representing the noisy filter. Both filters have pole position $z = K$. Box E computes the current noise sample generated by the filter of box D by subtracting from its output the presumably noiseless output of box C, and boxes F, G, and H compute the instantaneous mean-squared value of the noise.

Apart from any consideration of the validity of the program, we can see that it is characterized by two main features. One of these features is the smooth flow of control from routine to routine, and the other is the relative simplicity of the interactions of the various routines. By the first feature, we mean that, except for the initialization of register t, each routine is executed only once as control proceeds around the main loop, and the order of execution is the same forever. This is in marked distinction to the behavior of the control of many sorts of programs, where the flow chart is needed to describe the myriad possibilities for the order of execution of the various routines. But the smooth flow of control is the rule in next-state simulation programs. Since the topology of the flow chart is predictable, the flow chart gives little, if any, information and so loses much of its value as a program description.

The second feature of the program described by Fig. 5.2a, and by other next-state programs, involves the manner in which the routines interact. Each routine changes the contents of only a few registers (in this case, no routine changes more than one register), and these registers are changed by no other routines. Since each routine is executed once per pass of control around the main loop, each of these registers is changed only once per pass, and the register can be thought of as holding successive samples of a time waveform. Each time that control passes from box I to box A in the flow chart, all these registers have been updated; that is, they all contain the samples of certain waveforms at the same, tth, sampling instant. This means that, if we think of the *routines* as representing *elements* in a system, the *registers* they are permitted to change can be thought of as the *output nodes* of those elements. The registers that a routine looks at, but does not change, may be thought of as input nodes if they are changed by other routines, or as parameters of the element if they are fixed. This point of view permits representation of a next-state program by a block diagram such as Fig. 5.2b. This alternative program representation has the advantage of making evident the signal flow of the program, although it would be totally inadequate to describe the flow of control from routine to routine if the latter were at all

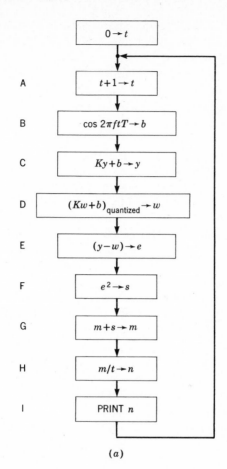

$$0 \rightarrow t$$

A $\quad t+1 \rightarrow t$

B $\quad \cos 2\pi f t T \rightarrow b$

C $\quad Ky+b \rightarrow y$

D $\quad (Kw+b)_{\text{quantized}} \rightarrow w$

E $\quad (y-w) \rightarrow e$

F $\quad e^2 \rightarrow s$

G $\quad m+s \rightarrow m$

H $\quad m/t \rightarrow n$

I $\quad \text{PRINT } n$

(a)

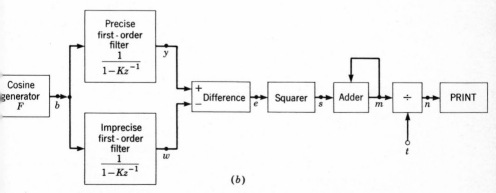

(b)

Fig. 5.2 (a) Flow-chart description of program to measure roundoff noise of digital filter; (b) block-diagram representation of same program.

intricate. The block-diagram representation also makes clear which routines are affected by which registers, and which registers are affected by which routines. It is a basic premise of this chapter that the signal flow is of much more interest than the flow of control for the sort of problems for which next-state techniques apply.

Once a block diagram has been drawn to represent the program, it is clear what is meant by the object system, namely, the system that would be described by the same block diagram. It is also reasonable to regard the set of registers that can be changed by the various routines as together comprising the state of the object system. Some of these registers correspond to *nodes* in the block diagram; others may be internal to some of the blocks. In Fig. 5.2b none of the state registers are internal to any of the blocks. However, if, for example, a second-order digital filter were part of the system, it would clearly need some storage registers for its previous inputs and outputs which would have to be updated each time the output is computed; these are internal-state registers. It is seldom important to count state registers exactly, but, for purposes of program debugging, it is useful to keep in mind just which registers are supposed to be changed and which routines are supposed to change them.

The next-state simulation program we have described has no way of coming to an end. In actual programming practice, unlike the real world being simulated, a program usually must be able to detect its own completion and yield control to another program. Some such completion-detection routine would appear in the flow chart but is not generally shown in the block diagram, except perhaps as a written comment. By the same consideration, a change in the completion routine is all that would usually be necessary to make a next-state program operate on longer or shorter records of data. In fact, the program can run for as long as it is supplied with input data or as long as its output data can be accepted. Therefore, if the program could run in real time, it would, in fact, be a realization of the object system. This has been achieved in the simulation of some relatively simple systems.

5.3 COMPARISON OF SOURCE AND OBJECT SYSTEMS

A next-state program is equivalent to an object system. If the program has no "bugs," the object system is well defined and understood. We shall examine some of the programming considerations later. In this section we are interested in how well or badly an object system represents a desired source system.

Examining the block diagram of the object system, we can compare it with the block diagram of the source system. Usually the genesis of the object system is the block diagram of the source system, and so the two block diagrams compare well in topology and in the functions of the blocks.   If, for example, the source system is a noise generator feeding a low-pass analog filter, the object system will naturally be chosen to be a discrete random-number generator feeding a digital low-pass filter.

Since one can almost always be assured of gross similarity of the source and object systems, the considerations that remain are usually those of sampling and quantization.   Of the two, sampling seems to be more important.   This is not surprising, because computers typically have a basic register length that permits better than 0.01 percent accuracy in basic arithmetic.†   This is about the state of the art of the best analog components, but it is more nearly a lower limit of computer accuracy since 24-bit and 32-bit machines are very common, and even a small machine can be programmed with multiple precision to obtain any desired arithmetic precision.   Nevertheless, accuracy must be considered, as we have seen in Chap. 4 in regard to filters.   Often a desired parameter is a very sensitive function of a stored quantity, as we found could be the case of the poles of a filter as a function of the coefficients of a difference equation.   Accuracy also enters into a system's behavior in the opposite way; a system will sometimes not work if it is too accurate.   For example, in Fig. 5.2a one of the elements generates a cosine wave of radian frequency $2\pi f$ at a sampling rate of $1/T$.   But if fT is any rational number the generated cosine wave will be truly, rather than approximately, periodic. If this is true, the digital filters may well reach a steady-state condition so that the measurements of quantization noise are repeated over and over again on the same data.   This would not very likely be what was intended, the cosine wave having probably been chosen as a random sort of input typical of the sorts of inputs to be processed by such filters. Another sort of difficulty arises when a certain input value is considered to have zero probability and therefore to be unimportant; in practice it may have a finite probability due to quantization.   The best-known case of this is the precaution against division by zero.

The inaccuracies due to sampling, if the source is analog rather than discrete, are very common, very necessary, and often very difficult to overcome.   The necessarily discrete object system can be only an approximation to an analog system, and the approximation cannot even be reasonable unless the waveforms *in all parts of the system* are adequately sampled.   The meaning of the word "adequate" may be subject to some engineering considerations, of course, but there is always a conflict between the desire for a high sampling rate for accuracy and a low sampling rate for speed of computation.   For band-limited waveforms the Nyquist rate is adequate, but most waveforms are not truly band-limited.

† A 14-bit register length satisfies this accuracy requirement.

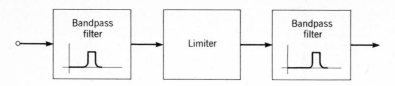

Fig. 5.3 Bandpass limiter.

As a case in point, let the source system be as shown in Fig. 5.3, a common subsystem in communications systems called a bandpass limiter. The limiter is a circuit whose output has a constant magnitude and the same sign as the input. To be specific, let both bandpass filters have bandwidths of 200 Hz centered around a 2,500-Hz center frequency, and let the input be a 2,500-Hz sine wave. Then the output of the limiter is a 2,500-Hz square wave, and the output of the second bandpass filter will be a 2,500-Hz sine wave. It is clear that the output amplitude is independent of the input amplitude because of the limiter. In a 200-Hz band around 2,500 Hz, the output amplitude is also independent of the input frequency.

Let us now consider an object system corresponding to the bandpass filter. Suppose the sampling rate for the object system is 10 kHz, a rate quite sufficient to describe the waveforms at the outputs of either of the bandpass filters. The waveform at the limiter output, however, does not have a band-limited spectrum but has frequency components at $2,500(2n - 1)$ Hz, where n goes from 1 to infinity. Therefore, in the object system, the sampled waveform at the limiter output will suffer from aliasing. The analog limiter output has a spectrum as shown in Fig. 5.4, and since aliasing consists in adding together all possible spectra resulting from shifting the original spectrum by multiples of 10 kHz, we see that the limiter in the object system will have a spectrum where the

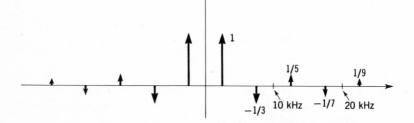

Fig. 5.4 Spectrum of output of bandpass limiter with 2,500-Hz sine-wave input.

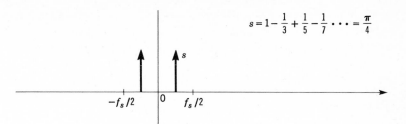

$$s = 1 - \frac{1}{3} + \frac{1}{5} - \frac{1}{7} \cdots = \frac{\pi}{4}$$

Fig. 5.5 Spectrum of output of object system corresponding to Fig. 5.3.

ines at 7,500, 12,500, 17,500 Hz, etc., will all have been folded into the region of the main line at 2,500 Hz, resulting in the spectrum shown in Fig. 5.5. For an input sine wave of exactly one-fourth of the sampling frequency these aliased lines actually coincide to give a single spectral line at 2,500 Hz, with magnitude of $\pi/4$ times what the analog situation produced. This means a distortion of 21.5 percent. If the input sine wave were not exactly 2,500 Hz, the aliased lines would not exactly coincide, so that in the neighborhood of 2,500 Hz the limiter output in the object system would have a number of distortion lines with amplitudes of $\frac{1}{3}$, $\frac{1}{5}$, $\frac{1}{7}$, etc., as well as the desired spectral line. If only the first of these distortion lines passed through the final bandpass filter, the distortion in its output would be 33 percent. It is easy to see that the distortion is a very sensitive function of the input frequency, the sampling frequency, and the filter passband.

Very similar situations will be encountered in the simulation of systems containing other nonlinearities, i.e., full-wave or half-wave rectifiers, logarithmic amplifiers, even squarers. In the last case, it is possible to argue that the highest frequency component in the output of the squarer is twice the highest frequency component in the input so that an overall sampling frequency of twice the Nyquist rate for the squarer input will permit a good simulation of the system. This is especially true if the squarer is to be followed by a low-pass filter, so that any folding effects involving only frequencies remote from zero frequency may be unimportant.

Generally, the solution of these distortion problems is to sample at a higher rate, but occasionally it may help to sample at a lower rate. In the above discussion of the bandpass limiter, a sampling rate of only 9 kHz would cause the largest distortion lines to fall at least 1 kHz away from 2,500 Hz so that they would easily be removed by the output bandpass filter. The limiter spectrum contains lines at $2,500(2n - 1)$ Hz, for all possible integers n. Only the two lines at $\pm 2,500$ Hz are desired.

Sampling at 9 kHz causes lines to appear at frequencies $[2,500(2n - 1) + 9,000m]$ Hz, for all n and m. We ask first which lines coincide with 2,500 Hz. For these lines, m and n must satisfy the equation

$$2,500 = 2,500 \times (2n - 1) + 9,000m \qquad (5.1)$$

which reduces to

$$n = 1 + \tfrac{9}{5}m$$

An obvious solution is $m = 0$, $n = 1$ corresponding to the desired line at 2,500 Hz. However, all m divisible by 5 will yield solutions. $m = 5$ yields a distortion line caused by the harmonic originally at $19 \times 2,500$ Hz; this line has a spectral value of $-\tfrac{1}{19}$. $m = -5$ causes a distortion line caused by the harmonic originally at $-17 \times 2,500$ Hz; this line contributes a distortion of $\tfrac{1}{17}$. $m = 10$ contributes a distortion line caused by the harmonic originally at $37 \times 2,500$ Hz; this line contributes a distortion of $\tfrac{1}{37}$. $m = -10$ contributes a distortion of $-\tfrac{1}{35}$, etc. By symmetry considerations, the desired line at $-2,500$ Hz has similar distortion terms. Since all lines in the limiter output spectrum are multiples of 1 kHz apart and are shifted by aliasing by multiples of 9 kHz, all distortion terms that are not solutions to (5.1) must lie more than 1 kHz distant from the desired lines. This fortuitous circumstance is because the greatest common factor of the sampling rate (9 kHz) and the line spacing (1 kHz) is so large. In the general case, many more distortion lines may need to be considered.

One of the important aspects of the comparison of source and object systems involves examining waveforms generated by the object system. For this purpose a sampling frequency of only twice the highest frequency present in the waveform is not adequate because, although it is mathematically possible to recreate the waveform from its Nyquist samples, it is very difficult to do this visually. This effect is illustrated by Fig. 5.6, which shows a sine wave and its samples taken at approximately one, two, four, and eight times the Nyquist rate. Figure 5.7 shows the same effect for a slightly more complicated waveform.

At low sampling rates a waveform can be interpreted more easily than otherwise if the time scale is expanded, or if the amplitude scale is compressed, so that points that should be "connected" are nearer than points that should not.

On the other hand, some non-band-limited waveforms may be effectively displayed, e.g., half-wave rectified waves. Also, for some kinds of waveforms, such as white noise, the spatter of dots resulting from sampling presents an impression that is satisfactory since the original waveform is not thought of as other than a fuzzy signal anyway.

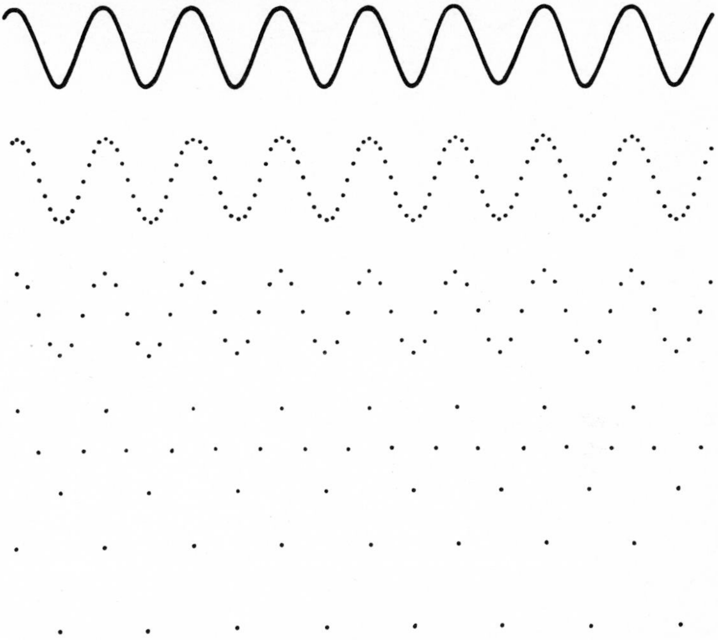

Fig. 5.6 Sine wave, together with its samples taken at one, two, four, and eight times the Nyquist rate.

5.4 ELEMENTS FOR OBJECT SYSTEMS

It is natural and sensible to build up a simulation program from a set of flexible and efficient subroutines, each of which represents an element likely to be encountered in many object systems and/or many times in the same object system. For example, linear filters are simulated as cascades of simple transfer functions such as

$$\frac{1}{1 - K_1 z^{-1}} \qquad \frac{1 - K_1 z^{-1}}{1 - K_2 z^{-1}} \qquad \frac{1 - K_1 z^{-1}}{1 - K_2 z^{-1} + K_3 z^{-2}}$$

where K_1, K_2, K_3 are parameters and where the simple transfer functions are realized in well-written general-purpose subroutines.   It is natural to try to catalog and classify such element-subroutine pairs.   One could ask of an element whether it has memory or no memory (the output of a no-memory element depends only on its present inputs) and whether it has no inputs, one input or several, or no outputs, one output or several. One may also ask whether an element has parameters.

The digital filter is a prime example of a very general and important

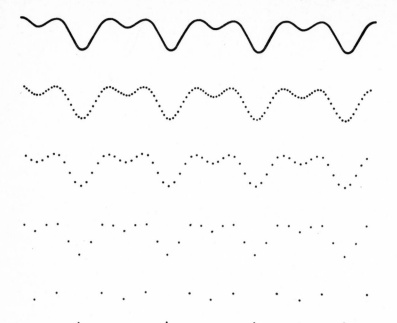

Fig. 5.7 Sum of two sine waves, together with samples taken at one, two, four, and eight times the Nyquist rate.

linear component, with memory and parameters and with one input and one output.   In distinction, there are linear elements with no memory, such as adders and attenuators, gains and inverters.   The adder is an example of a linear device with more than one input.   These devices are of an elementary nature, and the means of simulating them on a digital computer need not be discussed.   However, it is convenient to program subroutines to simulate them and to use these as components in a block diagram and the subroutines as components of a computer simulation.

Examples of simple nonlinear components with no memory are rectifiers—full-wave and half-wave—limiters, gates, switches, the "greater of" element, multipliers and dividers, logarithmic amplifiers, and square-root takers.   Many of these devices require only one or two computer instructions per sampling instant, and others can be efficiently programmed by borrowing from the literature on approximation of functions [3].

A simple component requiring memory is the peak detector.   If the peak of a sampled waveform is defined as the sample that is larger than both preceding and following samples, it is clear that both memory and a decision-making algorithm are needed.   A convenient way to represent peak detection is as shown in Fig. 5.8.

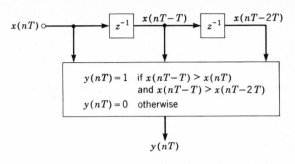

Fig. 5.8 Representation of peak detection in object system.

GENERATORS

Aside from linear and nonlinear devices, it is often desired that a system contain generators of various sorts, for example, pulse generators, sine and cosine generators, random-noise generators, square-wave generators, etc.† In fact, the programmer might consider that laboratory tests on new electrical devices are traditionally carried out by connecting, say, a signal generator to the input of the device and a "scope" or meter to various points of interest. Similarly, to test a simulated "device" may require that the output of a simulated generator be "connected" to the input of the "device" and a computer-controlled "scope" or printer monitor the interesting nodes in the simulated system.

RAMP, SAWTOOTH, TRIANGULAR-WAVE, AND SQUARE-WAVE GENERATORS

A ramp generator is an element whose output is of the form

$$y(nT) = An + B \tag{5.2}$$

where A and B are constants. It is simulated by letting $y(0) = B$ and performing the iteration

$$y(nT + T) = A + y(nT) \tag{5.3}$$

once per sampling interval.

A sawtooth generator is an element whose output is of the form of Fig. 5.9. If the iteration of (5.3) is performed long enough with large enough A, overflows will occur, causing $y(nT)$ to exhibit large negative jumps, as in Fig. 5.9. These jumps will occur once every M/A samples, where M is the difference between the largest positive and largest negative numbers that can be stored in a register. Thus (5.3) can provide a sawtooth generator where the sawtooth frequency is A/M. Full-wave

† Note that a stored table of sampled input to be fed into a next-state simulation is logically equivalent to having a generator, which we can call INPUT.

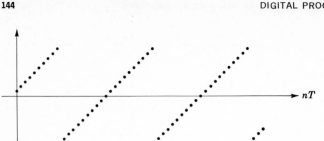

Fig. 5.9 Output of discrete sawtooth generator.

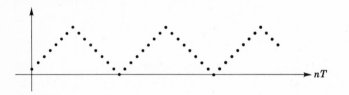

Fig. 5.10 Output of discrete triangular-wave generator.

rectification of the sawtooth-generator output produces the triangular wave shown in Fig. 5.10, and limiting (infinite clipping) the sawtooth-generator output produces the square wave shown in Fig. 5.11.

RANDOM-NUMBER GENERATOR

The generation of random numbers, as a representation of random noise, is usually achieved by a congruential scheme [5],

$$x_n = (Ax_{n-1}) \text{ modulo } P \tag{5.4}$$

where P is a large prime. By proper choice of A, this iteration can generate all the integers $1, 2, \ldots, P - 1$ in a seemingly random order

Fig. 5.11 Output of discrete square-wave generator.

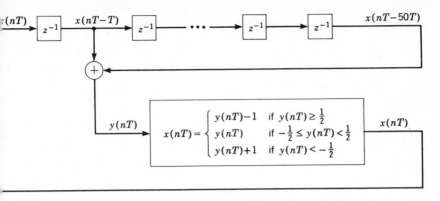

Fig. 5.12 Discrete random-number generator.

before repeating.   A disadvantage of (5.4) is the need for multiplication and, perhaps, division.   Another useful scheme, involving only addition, is

$$x_n = (x_{n-1} + x_{n-k}) \text{ modulo } 1 \tag{5.5}$$

or its equivalent [2], where k is a moderately large integer—say 50—and the initial k numbers are uniformly distributed between 0 and +1.   A very useful variation of (5.5), as diagrammed in Fig. 5.12, generates random numbers distributed uniformly between $-\frac{1}{2}$ and $+\frac{1}{2}$, and the *modulo* operation is achieved by the simple expedient of *ignoring overflow* in the fixed-point addition.

Equations (5.4) and (5.5) generate uniformly distributed noise samples, whereas in many simulation experiments gaussian noise is desired.   If x_n is uniformly distributed from 0 to 1, a two-step procedure for creating gaussian noise is as follows:

1. Compute a random variable y_n with a Rayleigh distribution (density)

$$P(y) = \frac{y}{\sigma^2} e^{-(y^2/2\sigma^2)} \tag{5.6}$$

using

$$y_n = \sqrt{2\sigma^2 \ln \frac{1}{x_n}} \tag{5.7}$$

2. Generate a pair of gaussian random numbers, w_n and w_{n+1}, with density

$$P(w) = \frac{1}{\sqrt{2\pi}\sigma} e^{-(w^2/2\sigma^2)} \tag{5.8}$$

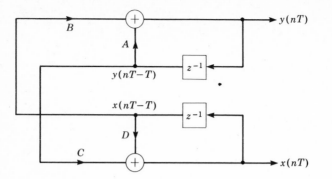

Fig. 5.13 Discrete sine-cosine generator.

using

$$w_n = y_n \cos 2\pi x_{n+1}$$
$$w_{n+1} = y_n \sin 2\pi x_{n+1}$$

(5.9

Two "independent" random numbers are generated at once. I
Rayleigh noise is desired, the second step of (5.9) is omitted.

SINE AND COSINE GENERATOR

A very important block in many simulation programs is a programme
sine-cosine generator, whose output can serve as a basic test signal fo
almost any system being simulated. Because of its importance an
because its digital realization is of theoretical interest, we shall study thi
generator in some detail.

First, consider the simultaneous difference equations

$$y(nT + T) = Ay(nT) + Bx(nT)$$
$$x(nT + T) = Cy(nT) + Dx(nT)$$

(5.10

The corresponding network is shown in Fig. 5.13.

Given the initial conditions $x(0)$ and $y(0)$, the z-transforms $X(z)$ o
$x(nT)$ and $Y(z)$ of $y(nT)$ can be derived. The result is

$$Y(z) = \frac{z^2 y(0) - Dzy(0) + Bzx(0)}{z^2 - z(A + D) + AD - BC}$$
$$X(z) = \frac{z^2 x(0) - Azx(0) + Czy(0)}{z^2 - z(A + D) + AD - BC}$$

(5.11

If the two conditions $AD - BC = 1$ and $A + D < 2$ are simultaneousl
realized, it is clear from (5.11) that the poles of the z-transforms must b
on the unit circle. This suggests that the outputs $x(nT)$ and $y(nT)$ ma

be sustained oscillations and, in fact, it can be demonstrated that the specific further conditions $y(0) = 0$, $x(0) = 1$, $A = D = \cos bT$, $B = -C = \sin bT$ yield the solutions

$$\begin{aligned} y(nT) &= \sin nbT \\ x(nT) &= \cos nbT \end{aligned} \tag{5.12}$$

NOISE CONSIDERATIONS IN SINE–COSINE GENERATOR

Equations (5.12) show that the network of Fig. 5.13 can be made into a sine-cosine generator by appropriately choosing the parameters.   At this point, it should be remembered that the quantization-noise results obtained in Chap. 4, specifically (4.20) and (4.17), state that the mean-squared output noise tends to infinity as the distance ϵ of the poles from the unit circle approaches zero.   The implication is clear, then, that the output of our sine-cosine generator eventually will become very noisy and, therefore, unusable.   However, (4.9) tells us that the mean-squared noise *gradually* increases, and if the rate of increase is slow enough so that many output cycles can be generated before appreciable noise develops, it becomes feasible intermittently to *reset* the variables so that the noise is effectively wiped out.

The noise formula may be derived by injecting independent round-off noise samples $e_1(nT)$ and $e_2(nT)$ at the two adders in the network of Fig. 5.13.   The z-transform of $x(nT)$ [keeping $y(0) = 0$, $x(0) = 1$, $A = D = \cos bT$, $B = -C = \sin bT$] then becomes

$$X(z) = \frac{z^2 - z \cos bT + zE_2(z) - \cos bTE_2(z) - \sin bTE_1(z)}{z^2 - 2z \cos bT + 1} \tag{5.13}$$

where $E_1(z)$ and $E_2(z)$ are the z-transforms of $e_1(nT)$ and $e_2(nT)$.

We see that the first two terms of the numerator correspond to the signal and the remaining terms to the noise.   Let us define

$$\begin{aligned} h_2(nT) &= Z^{-1}\left[\frac{z - \cos bT}{z^2 - 2z \cos bT + 1}\right] \\ h_1(nT) &= Z^{-1}\left[\frac{\sin bT}{z^2 - 2z \cos bT + 1}\right] \end{aligned} \tag{5.14}$$

where Z^{-1} is the inverse z-transform.   We can now use the basic equation (4.9) to obtain a formula for the total noise output, given the assumption that $e_1(nT)$ and $e_2(nT)$ are independent.

$$\sigma^2 = \frac{E_0{}^2}{12}\left[\sum_{m=0}^{n} h_1{}^2(mT) + \sum_{m=0}^{n} h_2{}^2(mT)\right] \tag{5.15}$$

Inspection of (5.14) shows that $h_1(nT)$ is actually sin nbT and $h_2(nT)$ i cos nbT. Substituting into (5.15) immediately yields the result

$$\sigma^2 = \frac{E_0^2}{12} n \qquad (5.16$$

Equation (5.16) states that the noise "power" increases linearly with the number of iterations of the difference equations. For example, afte 10^6 iterations, the standard deviation of the noise would correspond t about 9 bits. If a single iteration is performed in 100 μsec, severa minutes could elapse in an 18-bit machine before the generated sine an cosine waves become noisy.

RESETTING SINE–COSINE GENERATOR TO REDUCE NOISE

Returning to (5.11), if we impose the conditions $A = D = \cos bT$ $B = -C = \sin bT$ but let $x(0)$ and $y(0)$ be arbitrary, we obtain

$$y(nT) = x(0) \sin nbT + y(0) \cos nbT$$
$$x(nT) = -y(0) \sin nbT + x(0) \cos nbT \qquad (5.17$$

Equation (5.17) shows that $y(nT)$ and $x(nT)$ will both be sinusoida oscillations that are always in exact phase quadrature. Furthermore, i quantization effects are ignored, then, for any time nT, the equality

$$x^2(nT) + y^2(nT) = x^2(0) + y^2(0) \qquad (5.18$$

holds. However, because of noise, (5.18) will gradually cease to be tru as n increases.

In order to reset the system so that, after every k iterations, th variables $x(nT)$ and $y(nT)$ are changed to satisfy (5.18), we can multipl both $x(nT)$ and $y(nT)$ by the factor

$$f(nT) = \sqrt{\frac{x^2(0) + y^2(0)}{x^2(nT) + y^2(nT)}} \qquad (5.19$$

Thus, for each k iterations of (5.10), we perform once the nonlinea iteration

$$y(nT + T) = f(nT)[Ay(nT) + Bx(nT)]$$
$$y(nT + T) = f(nT)[Cy(nT) + Dx(nT)] \qquad (5.20$$

Execution of (5.20) effectively resets $x(nT + T)$ and $y(nT + T)$ s that (5.18) is satisfied.† Thus, if $x(nT)$ and $y(nT)$ had both drifted b the same relative amount to a lower value, both would be raised, in on

---

† In a private communication, Kaiser has suggested that limit cycles, which ca occur in the execution of (5.10) with roundoff, can also maintain the desired constan sine-wave amplitude.

iteration cycle, to the value they would have had if no noise were present. If, however, $y(nT)$ had drifted up and $x(nT)$ down so that the sum of the squares satisfied (5.18), then (5.20) would have no effect. Thus drifts of phase are not compensated by (5.20).

From (5.16) we can, by setting $n = k$, estimate how much noise to expect in k iterations and use this fact to determine how often (5.10) should be replaced by (5.20).

DELAY

We shall see in Sec. 5.5 that the simulation of systems involving delay elements needs special consideration. The delay element itself is easy to simulate if the delay is a multiple of the sampling interval. For a delay of k samples, k registers, numbered 0, 1, . . . , $k - 1$, should be set aside. To implement

$$y_n = x_{n-k} \tag{5.21}$$

the procedure is to put the contents of register n mod k into the output register Y and put x_n into register n mod k. This procedure takes an amount of time independent of k. For very small k, another procedure may be faster, as when the delay is part of a digital filter.

5.5 ORDERING OF EXECUTION OF SUBROUTINES IN A NEXT-STATE SIMULATION

In the typical next-state simulation, as described in Sec. 5.2, the instructions that simulate any element in the block diagram are performed once each time around the main loop of the program. These instructions examine all the current inputs to the element and compute the current output of the element. It is obvious that, for validity, all the current inputs to the element must have been computed before the routine simulating that element can be executed, and no routine requiring the current output can be executed until that output has been computed. This means that, within the main loop, not all possible orderings of execution of the subroutines corresponding to the elements of the block diagram result in valid programs. Consider, for example, Fig. 5.14, in which the detailed nature of the elements in the object system is unspecified.

It is clear that the output of element I for $t = 0$ must be computed first. The output of either element II or element IV could then be computed, but not elements III, V, or VI. The possible orders of

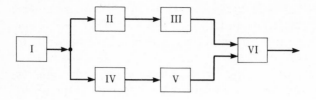

Fig. 5.14 System for which certain orders of execution
of subroutines are invalid.

execution of subroutines corresponding to the elements of Fig. 5.14 for
the first pass around the main loop of the program are listed below.

I, II, III, IV, V, VI
I, II, IV, III, V, VI
I, II, IV, V, III, VI
I, IV, V, II, III, VI
I, IV, II, V, III, VI
I, IV, II, III, V, VI

Any of the above orders of execution satisfies the requirement that all
inputs to a block at $t = 0$ have been computed before its subroutine has
been executed on the pass around the main loop that corresponds to
$t = 0$.   The same situation holds, in this case, for all subsequent passes
around the loop, corresponding to $t = nT$.

It is instructive to consider the effect of violating this requirement
by executing the subroutines in the order, say,

I, III, II, IV, V, VI

The first time around the loop, for $t = 0$, the output of block I is computed
and then subroutine III is executed.   But subroutine III expects to find
the current output of block II in some register, say R_{II}.   Instead it finds
some initial condition, and so the output of block III is incorrectly
computed.   Subroutine II is then executed, putting its correct output in
R_{II}, and subroutines IV and V are executed correctly.   Subroutine VI is
executed next but, since one of its inputs, from block III, was incorrectly
computed, the output of block VI is also incorrect.

On the second pass around the main loop, for $t = T$, subroutine I is
executed and then subroutine III.   This time, the expected input, in
register R_{II}, is the correctly computed value of block II from the previous
sampling instant.   In fact, on all subsequent passes, say the nth pass,
the output of element III will be based on the input from element II
computed on the $(n - 1)$st pass, so that the output of element III will
be correct except for a delay of T.   Similar reasoning shows that the

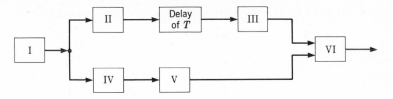

Fig. 5.15 System that is actually realized by invalid ordering of subroutines for system of Fig. 5.14.

outputs of I, II, IV, and V will be correct, and VI will be wrong because one of its inputs will be delayed with respect to the other. In fact, this may be illustrated by Fig. 5.15, a block diagram equivalent to what is actually realized by the program just described. Note that a delay has been inserted between elements II and III. An improper sequence of execution of subroutines in the object program always results in the insertion of invisible delay elements into the object system.

The system of Fig. 5.16, in which there is a simple feedback path, would seem to be impossible to simulate by the technique proposed because subroutine II cannot be executed properly until the output of III is known, and subroutine III cannot be executed properly until the output of II is known. In fact, the object system of Fig. 5.16 may well be ambiguous. If, for example, II is an adder with an output $y(nT)$ and III is a gain of G with an output $x(nT)$, then, if the output of I at $t = nT$ is denoted by $w(nT)$, the equations determining $x(nT)$ and $y(nT)$ are

$$x(nT) - y(nT) = w(nT) \qquad x(nT) - Gy(nT) = 0$$

which is a set of inconsistent equations (with no possible solution) if $G = 1$. For other values of G there are unique solutions for $x(nT)$ and $y(nT)$ but these solutions could not be computed by separate adder and gain subroutines. The only possible way in which the output of III could be computed properly by an independent subroutine before the output of II is known would be if the current output of element III did not depend on its current input. A delay element has this property; other elements used in simulations do not have this property (except generators), and in this restricted sense it can be said that the delay element is the only element whose current output does not depend on its current input. Thus, if element III in Fig. 5.16 is a delay of NT, the

Fig. 5.16 Object system with feedback path; in general, this may not be simulated by next-state simulation.

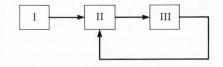

object program can be written by ordering the subroutines as

I, II, III$_A$

where III$_A$ is a subroutine that simulates a delay of $(N - 1)T$, which just compensates the delay of T that is inserted because of the execution of subroutine II before III$_A$ has been executed.

We shall now state a procedure for the ordering of routines in an object program.

I. Search the block diagram for an element whose subroutine has not been incorporated in the object program. If this element is a generator (or an INPUT) or if all its inputs are from elements whose subroutines have been incorporated in the object program already, then the subroutine for this element may be incorporated next in the program. If there are more elements satisfying this requirement repeat I; otherwise carry out instruction II.

II. If there are no more elements, the object program is complete and is a valid representation of the object system. If there are more elements but none of them is a delay element, then next-state simulation cannot validly represent the object system. If at least one of the elements remaining is a delay, proceed to III.

III. Choose a delay element whose subroutine has not been incorporated in the object program. If the delay time is supposed to be NT, incorporate a subroutine simulating a delay of $(N - 1)T$ next in the object program, and consider this delay element to have been represented. (The special case of a delay of zero should be considered carefully.) If more elements remain in the system, return to instruction I.

The failure condition detected by II implies that there is a closed (feedback) loop in the object system that has no delays; therefore, one must start somewhere in violation of the above rules. The best thing to do may be to modify the object system by the insertion of a delay of T and continue the procedure. One's engineering insights must be used to determine the degree of discrepancy between source and object systems caused by this delay. Physical systems, of course, must have some delay, though perhaps it may be much less than T.

5.6 THE PATSI COMPILER

Once one has decided that an object system with the same block diagram as the source system is a valid representation of the latter, programming a system simulation consists almost entirely in using the routines described in the preceding sections and connecting them together in the way described in Sec. 5.5.   Much of this is nuisance work, however, and with a simple and formalized description of the system to be simulated, this programming can be done by a compiler (a program that can write other programs).   The compiler could do the work of looking up the routines to be used and copying them into memory, allocating the various storage registers for the outputs of the several circuits in the system, computing the coefficients of the various difference equations, setting the initial conditions, etc.   Several such compilers have been written and used to date [4, 6].   One of them, called PATSI (an acronym for Programming Aid To System Investigation), will be described below.   It is simple enough that it can be programmed in a short time.   If an assembler that can accept macrodefinitions with parameters is available, PATSI can be programmed in just a few man-days of time.

Describing a compiler means first explaining the language with which a problem is described and, second, describing the implementation of a program that translates the problem description into a program.   We shall begin with a description of the PATSI language, using as an example the system of Fig. 5.17.   This is a system and its complete PATSI program.   The system is not meant to have any utility but is illustrative only.   The text below the diagram is the complete source program (the source program is the user's description of his problem).

In the block diagram we note the letters X, Y, W, Q, etc.   These

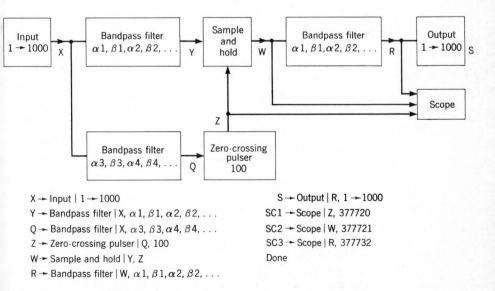

$X \rightarrow$ Input | $1 \rightarrow 1000$
$Y \rightarrow$ Bandpass filter | X, $\alpha1, \beta1, \alpha2, \beta2, \ldots$
$Q \rightarrow$ Bandpass filter | X, $\alpha3, \beta3, \alpha4, \beta4, \ldots$
$Z \rightarrow$ Zero-crossing pulser | Q, 100
$W \rightarrow$ Sample and hold | Y, Z
$R \rightarrow$ Bandpass filter | W, $\alpha1, \beta1, \alpha2, \beta2, \ldots$

$S \rightarrow$ Output | R, $1 \rightarrow 1000$
SC1 $\rightarrow$ Scope | Z, 377720
SC2 $\rightarrow$ Scope | W, 377721
SC3 $\rightarrow$ Scope | R, 377732
Done

Fig. 5.17 Source system and its complete description in PATSI language.

letters have several meanings. For example, W refers to the sample and-hold circuit, to its output waveform, to its output node, and to the line of the source program describing the preceding three entities. There is, as we shall see, no ambiguity in the quadruple use of the name. Le us examine line W more carefully. The name W has already been explained. It is separated from what follows by an arrow. The name assigned to the circuit is therefore the set of letters preceding the arrow it need not be a single letter or even a single word. SAMPLE ANI HOLD is a functional description of the circuit and is one of about 5(functional types that PATSI knows how to simulate. The vertical ba: separates the functional description from what follows, the meaning o which depends on the type of element. For SAMPLE AND HOLD the next two names are inputs, separated by commas. The first name is the name of the input to be sampled, Y, and the second name is the name of a pulse waveform to initiate the sampling, Z. The order Y, Z indicate which is which. To use the SAMPLE AND HOLD block in a system one must know this convention,† and for this purpose a dictionary of PATSI elements exists and is kept up to date.

Let us now look at line Z, describing a zero-crossing pulser, a block whose output is zero except when the input changes sign. Here the entities following the vertical bar are Q, the input, and 100, a parameter of the block whose meaning is the height of the pulse emitted. The meanings and order of these two quantities are also found in the dictionary

Line X describes an INPUT block. Functionally, INPUT is a generator, an element with one output and no inputs. The successive samples of the waveform at node X are taken from successive registers in memory, starting in this case with register 1 and ending with register 1000. INPUT permits a simulation program to be run with real data. Different modes permit INPUT to get data from bulk storage instead o core memory, permitting effectively infinite-length data.

Similarly, the OUTPUT block S describes an element that has no outputs but one input, in this case R. The successive samples of the waveform R are stored in successive designated registers in memory.

The three lines SC1, SC2, and SC3 describe the extremely useful SCOPE element. This element, like OUTPUT, has inputs but no outputs. Each time around the main loop the SCOPE routine displays the samples of its input waveforms on a digital oscilloscope, and the horizontal position is incremented to provide a "sweep." The second input is the address of a register used to control gain, vertical position, and intensity. The addresses used here are those of toggle-switch registers on the TX-2 computer, which may be hand set and varied while the program is running.

† Just as one would have to know the convention with a real sample-and-hold circuit

Note that the SCOPE display occurs while the program is running. Thus, since the object program is a slow-motion version of the object system, the effect is exactly as if one had the object system in the laboratory and were studying it with an oscilloscope. As many waveforms as desired can be observed together since there is no difficulty in attaching SCOPE to several nodes in the system. We also use an XYSCOPE, which permits Lissajous figures to be displayed in the same way or even at the same time as waveforms are displayed.

The SCOPE is an extremely powerful debugging tool, not only for debugging a simulation program but for debugging the system under consideration. Other graphical output features may also be incorporated into a simulation program in this way.

The line DONE simply means that the source program (description of the source system) is complete.

We are now ready to describe the action of the compiler in generating the object program. In the implementation involving only a macro-assembler, provided the proper ordering of the routines has been accomplished by reordering the lines of the source program, each line of the source program is recognized as a use of a predefined macrostatement with parameters.

The compiler first copies into memory a starting program. This is an extremely simple routine whose function is to jump to the first subroutine after setting an index register, say L, to point to the first output node. This node is always at a fixed location, called the start of the calling sequence. The location of the first subroutine to enter is always found in the second register in the calling sequence.

The compiler then looks at the first line of the source program. This line is

$$X \to \text{INPUT}|1 \to 1000$$

The name X is associated with the location of the first register of the calling sequence. This first register represents the output node of the block. In the next register is put the location of the beginning of the input subroutine. Note how this satisfies the conditions required by the starting routine. In the next two registers of the calling sequence are put the first and last locations of the input samples, 1 and 1000. During the operation of the simulation program, the subroutine will know to find these parameters in locations 2_L and 3_L.† In the next register is put the first input location again. This is a state register which will be updated by adding 1 after each execution of the input subroutine, thus pointing to the register containing the next sample. Thus the part of the calling

---

† The notation a_L should be read "a indexed by the contents of L."

sequence generated to correspond with the line X of the source program contains the output node, the location of the proper subroutine, the necessary parameters for the subroutine to use, and the necessary state register of the block, all of which can be located, using the index register L, by the proper subroutine.

The input subroutine, when it has finished its main job, increases the contents of index register L by 5, the number of registers in the calling sequence corresponding to X. It then looks in location 1_L to find the location of the beginning of the subroutine to simulate block Y, as we shall see.

After the compiler has dealt with the first line of the source program it looks at the next line,

$$Y \to \text{BANDPASS FILTER} | X, \alpha 1, \beta 1, \alpha 2, \beta 2, \dots$$

and associates the name Y with the next location in the calling sequence. In the next register is put the location of the bandpass-filter simulation subroutine. Note how this satisfies the conditions required by the input subroutine, which looked in register 1_L to find the location of the subroutine to which to go.

The location X is put in the next register in the calling sequence. The bandpass filter is able to find the location of its input by looking in register 2_L. In the next few registers are put the parameters of the filter $\alpha 1, \beta 1, \alpha 2, \beta 2, \dots$. The ellipsis here is not part of the program but a shorthand to indicate that many more parameters are typically used. The registers after these are state registers which will contain the previous values of input and output needed by the difference equation that simulates the filter. We shall deal with the calculation of coefficients of the difference equation later. During the simulation, the bandpass-filter subroutine can find all the parameters and state registers it needs by using index register L, and when it has finished its computation it will increase the contents of index register L by the length of its section of the calling sequence and find the location of the next subroutine to go to in 1_L.

All elements with outputs have the output node in the first register which corresponds to that element in the calling sequence, and all elements, whether with outputs or not, have the location of the simulation subroutine in the second register corresponding to that element. All simulation subroutines are written with the assumption that, when they are executed, index register L will be pointing to the first register in the appropriate part of the calling sequence, and all simulating subroutines finish by adding the length of this portion of the calling sequence to the contents of index register L. These simple conventions allow the subroutines to work together without unpredictable interferences with each other. It is these sections of calling sequence that are translated by macrodefinitions; the remainder of a simulation program, namely, the subroutines, can be written in ordinary assembly language once and for all.

The compiler creates more registers of the calling sequence for each of the other lines of the program in the same way. The line DONE is translated into a pair of registers in the calling sequence. The first register is meaningless, and the next is the location of the starting routine, which is always a fixed location.

The coefficients of difference equations, and any other parameters to be converted to more appropriate forms for the use of the simulation subroutines, are not computed by the compiler but by the subroutines themselves on the first execution of these routines called by a particular part of the calling sequence. Suppose a routine to simulate the Laplace transfer function $a/(s + a)$ is desired, using an impulse-invariance criterion. PATSI has such a block called RC POLE, and the calling sequence generated for it contains four registers,

output register
subroutine entrance
input location
$-aT$

although the subroutine will use the quantity e^{-aT} in the execution of the difference equation. To accomplish the transformation of the parameter to the more convenient form, the start of the RC POLE subroutine gets $-aT$ from register 3_L and computes e^{-aT}, which it stores in 3_L. It then modifies the calling-sequence-register 1_L (subroutine entrance) so that, on future passes through the program, the first part of the subroutine will not be executed but only the part meant to be done every time.

Two useful macros, BREAK and RETURN, ought to be part of the language of a PATSI compiler. These are not element-simulating statements, but they allow the introduction of machine-language programs between the simulation of elements in the object program. BREAK is translated into two registers in the calling sequence—the first being meaningless, and the second containing the location of the register immediately after it. Thus, when the previous subroutine is completed, it will look in 1_L and find that it is to transfer to register 2_L where it will find the machine-language program. RETURN produces a program much like the starting routine, and the calling sequence is resumed after the RETURN statement. These conventional programming insertions are useful for taking statistics during a simulation, providing options in the program, and simulating elements that will probably never be used again.

Another useful macro is MULTIPLY T BY n, which is like DONE except that it is executed only $n - 1$ out of every n times around the loop. Thus elements described after the MULTIPLY T BY n have an effective sampling rate of $1/n$ times the system rate, which makes the simulation program that much faster.

5.7 SUMMARY

Next-state simulation is a technique for simulating source systems. An object system is chosen that corresponds closely to the desired source system but is necessarily discrete. The object system is then simulated by an object program made up of subroutines, each of which, when entered, computes the "next" output of one element in the system. The object program can be described by a block diagram identical to the block diagram of the object system, which in turn will be very similar to the block diagram of the source system   The ordering of the subroutines in the object program is determined by the topology of the block diagram. It is possible to exploit the similarity of the source-system block diagram and the object-program block diagram by programming the simulation, using a higher-level language that is a description of a block diagram. This makes it possible to program simulations of complex systems in a minimum of time and thus encourages some research into the behavior of systems that might otherwise be too complicated to consider.

REFERENCES

1. Briscoe, H. W., and P. L. Fleck: A Real Time Computing System for LASA, presented at Spring Joint Computer Conference, *AFIPS Proc.*, **28**: 221–228 (1966).
2. Green, B. F., J. E. K. Smith, and L. Klem: Empirical Tests of an Additive Random Number Generator, *J. Assoc. Computer Machinery*, **6**(4): 527–537 (October, 1959).
3. Hastings, C., Jr.: "Approximations for Digital Computers," Princeton University Press, Princeton, N.J., 1955.
4. Kelly, J. L., Jr., C. L. Lochbaum, and V. A. Vyssotsky: A Block Diagram Compiler, *Bell System Tech. J.*, **40**: 669–676 (May, 1961).
5. MacLaren, M. D., and G. Marsaglia: Uniform Random Number Generators, *J. Assoc. Computer Machinery*, **12**: 83–89 (1965).
6. Rader, C. M.: Speech Compression Simulation Compiler, presented at Spring Meeting of Acoustical Society of America, 1965, *J. Acoust. Soc. Am.* (*A*), June, 1965.

6
Discrete Fourier Transforms

6.1 CONTINUOUS FOURIER–TRANSFORM THEORY

The Fourier transformation is one of the most important mathematical aids to signal processing and is of importance both in theory and in practice. It has wide applications outside the signal-processing field as well.

The Fourier transform, or spectrum, of a signal $f(t)$ is defined as

$$F(\omega) = \mathfrak{F}\{f(t)\} = \int_{-\infty}^{\infty} f(t)\, e^{-j\omega t}\, dt \tag{6.1}$$

Both $f(t)$ and $F(\omega)$ may be complex functions of a real variable. The basic property of the Fourier transform is its ability to distinguish waves of different frequencies that have been additively combined. Thus, if $\delta(x)$ is the unit-impulse "function,"

$$\mathfrak{F}\left\{\sum_i a_i e^{j\omega_i t}\right\} = 2\pi \sum_i a_i \delta(\omega - \omega_i) \tag{6.2}$$

that is, a sum of sine waves, overlapping in time, transforms into a sum of impulses which, by definition, are nonoverlapping. In signal-processing terminology, the Fourier transform is said to represent a signal in the frequency domain, and ω, the argument of the Fourier transform, is referred to as frequency.

It can be shown from (6.1) that the Fourier transform is *linear* since

$$\mathcal{F}\{x(t) + y(t)\} = \mathcal{F}\{x(t)\} + \mathcal{F}\{y(t)\} \tag{6.3}$$

and, for c a constant,

$$\mathcal{F}\{cx(t)\} = c\mathcal{F}\{x(t)\} \tag{6.4}$$

Finally, it is an invertible transform, with the inverse given by

$$f(t) = \mathcal{F}^{-1}\{F(\omega)\} = \frac{1}{2\pi} \int_{-\infty}^{\infty} F(\omega)e^{jt\omega} \, d\omega \tag{6.5}$$

The inverse Fourier transform (6.5) is remarkably similar in form to (6.1). For this reason, all transform "pairs" have near mirror images. Thus, the time function†

$$x(t) = \begin{cases} 1 & \text{for } |t| < T/2 \\ 0 & \text{for } |t| > T/2 \end{cases}$$

shown in Fig. 6.1a, has the Fourier transform

$$X(\omega) = T\frac{\sin{(\omega T/2)}}{\omega T/2}$$

shown in Fig. 6.1b, and the time function

$$y(t) = \frac{\omega_0}{2\pi}\frac{\sin{(\omega_0 t/2)}}{\omega_0 t/2}$$

shown in Fig. 6.1c has the transform

$$Y(\omega) = \begin{cases} 1 & \text{for } |\omega| < \omega_0/2 \\ 0 & \text{for } |\omega| > \omega_0/2 \end{cases}$$

shown in Fig. 6.1d.

Historically, the Fourier transform was developed from the notion of Fourier series. However, the latter can be shown to be a special case of the former. If $f(t)$ is periodic, the Fourier transform $F(\omega)$ consists of a weighted sum of impulses at frequencies that are integer multiples of the fundamental frequency of $f(t)$, and the weights associated with the impulses are the Fourier series coefficients associated with one period of $f(t)$.

---

† Since the Fourier transform is defined by an integral, the ambiguity in the definition of $x(t)$ at $\pm T/2$ does not affect the transform.

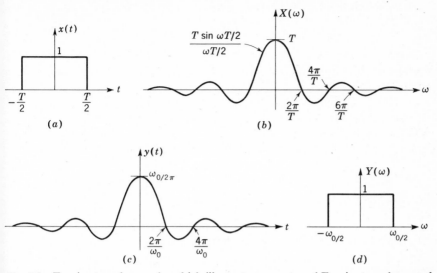

Fig. 6.1 Fourier transform pairs which illustrate symmetry of Fourier transform and inverse Fourier transform.

The Fourier transform is very closely related to the Laplace transform. For time functions that exist only for $t \geq 0$, the Laplace transform is defined by

$$\mathcal{L}\{f(t)\} = \int_0^\infty f(t)e^{-st}\,dt \qquad \text{a function of } s \tag{6.6}$$

For $s = j\omega$, the Laplace transform equals the Fourier transform. Conversely, computing the Fourier transform of $f(t)e^{-\alpha t}$ gives the Laplace transform of $f(t)$ for values of $s = \alpha + j\omega$, and so the Fourier transform may be used to compute the Laplace transform.

It is meaningless to talk about the Fourier transform of a sequence of samples. However, we can associate a discrete sequence of numbers $f(nT)$ with the time function

$$f^*(t) = \sum_{n=-\infty}^{\infty} f(nT)\,\delta(t - nT) \tag{6.7}$$

which has the Fourier transform

$$F^*(\omega) = \sum_{n=-\infty}^{\infty} f(nT)e^{-j\omega nT} \tag{6.8}$$

If the sequence is zero for $n < 0$, it has a z-transform

$$F(z) = \sum_{n=0}^{\infty} f(nT)z^{-n}$$

and we see that

$$F^*(\omega) = F(e^{j\omega T}) \tag{6.9}$$

We see also that any time function of the form of (6.7), a sequence of equally spaced impulses, has a transform that is periodic in frequency with period $2\pi/T$.

One final property of the Fourier transform is important, the convolution property; namely, for any $x(t)$ and $y(t)$ with Fourier transforms $X(\omega)$ and $Y(\omega)$, the function

$$\int_{-\infty}^{\infty} x(\tau)y(t - \tau)\,d\tau$$

has the Fourier transform $X(\omega)Y(\omega)$.

6.2 THE DISCRETE FOURIER TRANSFORM

For many reasons, we should like to be able to compute Fourier transforms with digital machines. This means we must consider only discrete samples of both the time function and the spectrum and only a finite number of samples of each. Suppose the time function $f(t)$ is represented by the sequence of N samples $f(nT)$, $0 \leq n \leq N - 1$, where T is the sampling interval in the time domain. Then, similarly, let the spectrum $F(\omega)$ be represented by $F(k\Omega)$, $0 \leq k \leq N - 1$, where Ω is the chosen increment between samples in the frequency domain. If, from the sequence $f(nT)$, we form the function $f^*(t)$ of (6.7), with a finite number of impulses, as shown in Fig. 6.2, we can write an expression for $F^*(k\Omega)$,

$$F^*(k\Omega) = \sum_{n=0}^{N-1} f(nT)e^{-j\Omega Tnk} \tag{6.10}$$

which is completely discrete in both time and frequency and is therefore suitable for computation.

With the above as motivation, we define the discrete Fourier transform, abbreviated DFT, of a sequence of N samples, $f(nT)$, $0 \leq n \leq N - 1$, as another sequence,†

$$F(k\Omega) = \sum_{n=0}^{N-1} f(nT)e^{-j\Omega Tnk} \tag{6.11}$$

where

$$\Omega = \frac{2\pi}{NT} \tag{6.12}$$

† Unless otherwise indicated, we shall assume both $f(nT)$ and $F(k\Omega)$ are complex.

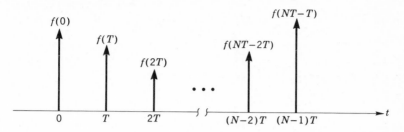

Fig. 6.2 Time function which can be associated with sequence of samples.

With this specification of Ω, there are only N distinct values computable by (6.11), namely, those for $k = 0, 1, \ldots, N - 1$. To prove this, we consider (6.11) for an arbitrary integer k, expressed as

$$k = rN + k_0$$

where $k_0 = k$ modulo N, and $r = (k - k_0)/N$. Then

$$F(k\Omega) = \sum_{n=0}^{N-1} f(nT)e^{-j\Omega Tn(rN+k_0)} \tag{6.13}$$

but since $\Omega TN = 2\pi$ the last expression can be reduced to

$$F(k\Omega) = F(k_0\Omega) \tag{6.14}$$

We may think of (6.11) as a formula yielding a sequence of only N numbers or as a formula yielding a periodic sequence of numbers with period N. *Alternatively, the discrete Fourier transform may be thought of as an evaluation of the z-transform of the finite sequence $f(nT)$ at N points in the z plane, all equally spaced along the unit circle at angles of $k\Omega$ radians.* Considering the sequence of spectral samples as arranged around a circle, as in Fig. 6.3, instead of along a line, is very helpful in certain situations. The spectrum of an analog function composed of equally spaced impulses could be similarly conceived as plotted around a circle since it is periodic in frequency.

In the remainder of this section we shall consider the properties of the discrete Fourier transform in some detail. These properties parallel those of the continuous Fourier transform. It is important to realize that, although a DFT of a sequence can only approximate the Fourier transform of a function, the properties of the DFT are *exact* properties rather than approximate properties based on the analog Fourier transform's properties [5, 9]. In what follows, the expression x modulo N will occur frequently, often in complicated expressions. We therefore adopt a shorthand notation:

$$((x)) = x \text{ modulo } N = x + rN \qquad 0 \le x + rN < N \tag{6.15}$$

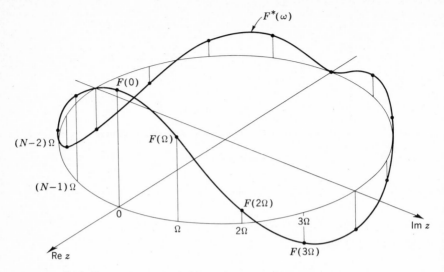

Fig. 6.3 Periodic spectrum plotted in z plane around unit circle.

We shall often make use of the fact that

$$e^{-j\Omega T((x))} = e^{-j\Omega T x}$$

The discrete Fourier transform of a complex sinusoid of frequency $q\Omega$,

$$f(nT) = e^{jq\Omega nT} \tag{6.16}$$

may be deduced from the definition (6.11) to be

$$F(k\Omega) = \sum_{n=0}^{N-1} \nu^n = \frac{1 - \nu^N}{1 - \nu} \tag{6.17}$$

where $\nu = e^{j\Omega T(q-k)}$.   This holds whether or not q is an integer.   In the former case, (6.17) may be further simplified.   If q is an integer, $F(k\Omega) = 0$ for $k \neq ((q))$ because $\nu^N = 1$.   For $k = ((q))$, the right side of (6.17) is of the form $0/0$ since $\nu = 1$, but the summation is easily seen to be equal to N.   Thus a complex sinusoid whose frequency is an integral multiple of Ω has the discrete Fourier transform

$$F(k\Omega) = \begin{cases} N & \text{for } k = ((q)) \\ 0 & \text{otherwise} \end{cases} \tag{6.18}$$

which implies that the frequency-selectivity property

$$\mathrm{DFT}\left\{\sum_q a_q e^{jq\Omega nT}\right\} = Na_k \tag{6.19}$$

of the discrete Fourier transform corresponds closely to the continuous Fourier-transform property of (6.2). However, there are only N distinct numbers in the DFT and so only N different frequencies can be distinguished perfectly. For frequencies that are not integer multiples of Ω we see, from (6.17), that all the points of the DFT are nonzero.

The linearity properties of the DFT,

$$\text{DFT } \{f(nT) + g(nT)\} = \text{DFT } \{f(nT)\} + \text{DFT } \{g(nT)\} \qquad (6.20)$$

and

$$\text{DFT } \{c(f(nT))\} = c[\text{DFT } \{f(nT)\}] \qquad (6.21)$$

corresponding to (6.3) and (6.4), also follow directly from the definition (6.11).

There exists an inverse DFT, a transformation that maps a discrete Fourier transform back into the sequence from which it was computed. It is given by

$$f(lT) = \frac{1}{N} \sum_{k=0}^{N-1} F(k\Omega)e^{+jTkl\Omega} \qquad (6.22)$$

which differs from (6.11) only by a scale factor and in the sign of the exponential, just as was the case with the analog Fourier transform's inverse (6.5). To prove that (6.22) is really an inverse, we substitute (6.11) into the sum of (6.22), which gives

$$\frac{1}{N} \sum_{k=0}^{N-1} \sum_{n=0}^{N-1} f(nT)e^{-j\Omega Tnk}e^{+j\Omega Tlk}$$

Reversing the order of summations, we can factor $f(nT)$ out of the sum over k, which gives

$$\frac{1}{N} \sum_{n=0}^{N-1} f(nT) \left(\sum_{k=0}^{N-1} e^{j\Omega T(l-n)k} \right)$$

where the sum over k is recognized as being of the same form as the sum of (6.17). The inner sum is thus equal to N for $((l)) = n$, and zero otherwise. Thus only one term in the outer sum is nonzero, and (6.22) is proved.

The inverse DFT, like the DFT, can produce samples $f(lT)$ for l outside the range $0 \leq l \leq N - 1$ but these are simply repetitions of the values of $f(lT)$ taken on with l within this range. Thus we can regard the values of $f(lT)$ as arranged around a circle, or as periodically extended, etc.

Let us now consider the inverse discrete Fourier transform of the product of discrete Fourier transforms, for which the corresponding analog result would be a convolution. That is, we are considering

$$v(lT) = \frac{1}{N} \sum_{k=0}^{N-1} X(k\Omega) Y(k\Omega) e^{jlTk\Omega} \tag{6.23}$$

where $X(k\Omega)$ and $Y(k\Omega)$ are the DFTs of sequences $x(nT)$, $y(nT)$, respectively. We assume, of course, that $x(nT)$ and $y(nT)$ each have an equal number of points, N. To evaluate (6.23) we substitute the definitions of $X(k\Omega)$ and $Y(k\Omega)$, which gives

$$v(lT) = \frac{1}{N} \sum_{k=0}^{N-1} \left[\sum_{n=0}^{N-1} x(nT) e^{-j\Omega Tnk} \right] \left[\sum_{m=0}^{N-1} y(mT) e^{-j\Omega Tmk} \right] e^{j\Omega Tlk}$$

$$\tag{6.24}$$

which we can write as a triple sum over indices k, m, and n. If we interchange the order of summations so that the innermost sum is over k, we can factor $x(nT)y(mT)$ out of this inner sum, which gives

$$v(lT) = \frac{1}{N} \sum_{m=0}^{N-1} \sum_{n=0}^{N-1} x(nT)y(mT) \left[\sum_{k=0}^{N-1} e^{j\Omega T(l-m-n)k} \right] \tag{6.25}$$

The bracketed sum is now of the form we have encountered in the two previous proofs and is zero for all m and n except combinations of m and n that satisfy

$$m = ((l - n)) \tag{6.26}$$

for which the bracketed sum is equal to N. Thus (6.25) reduces to

$$v_l = \sum_{n=0}^{N-1} x_n y_{((l-n))} = \sum_{n=0}^{N-1} y_n x_{((l-n))} \tag{6.27}$$

where we have switched to a subscripted notation for simplicity. We have thus shown that the product of discrete Fourier transforms is the discrete Fourier transform of a *circular* convolution [14]. The meaning of (6.27) is illustrated by Fig. 6.4. Here the five-point sequence $x(nT)$, written x_n, is circularly convolved with the five-point sequence y_m. We can choose to consider one or both sequences as periodically extended (white dots), or we can consider them as plotted around circles. Line 1 of Fig. 6.4 shows the sequence x_n, and line 2 shows the periodically extended sequence y_m. On line 3, representing the convolution sum v_0, the y sequence has been reversed in time and is multiplied by the corresponding member of the x sequence. Lines 4 to 7 show how the time-reversed y sequence is shifted for the computation of v_1, v_2, v_3, and v_4. As a sample of y_m is shifted out of range on the right side, the same sample is shifted into range from the left.

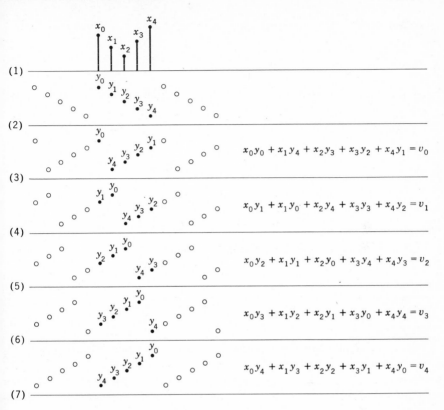

Fig. 6.4 Periodic, or circular, convolution.

Figure 6.4 illustrates why circular convolution is sometimes called *periodic* convolution. By contrast, the corresponding property of the continuous Fourier transform might be called *aperiodic* convolution. In most problems involving convolution, aperiodic convolution is meant.

A corollary of the circular-convolution theorem is the circular-shifting theorem

$$\text{DFT}\left\{f(((n-l))T)\right\} = F(k\Omega)e^{-j\Omega Tlk} \qquad (6.28)$$

which states that moving l samples from the end of a sequence to the beginning of the sequence is equivalent to multiplying the discrete Fourier transform by a linear-phase function. This is analogous to the continuous Fourier transform of a delayed-time function.

Some of the properties of the discrete Fourier transform are related to odd and even sequences. An odd sequence of N samples, $x(nT)$, is defined by

$$x(nT) = -x(((-n))T) \qquad (6.29)$$

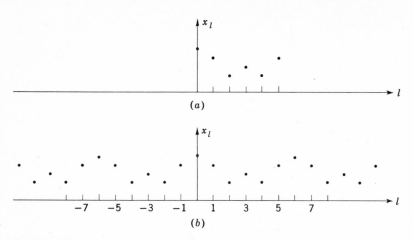

Fig. 6.5 (a) Even sequence; (b) its periodic extension.

and an even sequence of N samples, $y(nT)$, by

$$y(nT) = y(((-n))T) \qquad (6.30)$$

If the sequences are plotted around a circle, or periodically extended, these definitions correspond closely to the familiar definitions of odd or even functions. This is illustrated by Fig. 6.5a, which shows an even sequence of six samples, and Fig. 6.5b, which shows the periodic extension of Fig. 6.5a.

An even sequence with real samples has a discrete Fourier transform that is also real and even. An odd sequence with real samples has a discrete Fourier transform that is odd and pure imaginary. Any real sequence has a transform whose real part is even and whose imaginary part is odd. The transform of a sequence that is even but not purely real will also be even but not purely real, and the transform of a sequence that is odd but not pure imaginary will be odd but not pure real. These and other related properties are summarized in Table 6.1. By the symmetry of the definitions of the transform and its inverse, all the entries in the table remain valid if the headings of the columns $f(nT)$ and $F(k\Omega)$ are interchanged.

Let us now consider the properties of discrete Fourier transforms of artificially lengthened sequences. Thus, suppose we have samples $f(nT)$, $0 \leq n \leq N - 1$, and we create a longer sequence $g(nT)$, $0 \leq n \leq rN - 1$, where r is any integer and where

$$g(nT) = \begin{cases} f(nT) & 0 \leq n \leq N - 1 \\ 0 & \text{otherwise} \end{cases} \qquad (6.31)$$

In the definition of the DFT of $g(nT)$ we must use Ω/r in place of Ω.

$$G\left(k\left[\frac{\Omega}{r}\right]\right) = \sum_{n=0}^{rN-1} g(nT)e^{-j\Omega Tnk/r} = \sum_{n=0}^{N-1} f(nT)e^{-j\Omega Tnk/r} \qquad (6.32)$$

Thus, if k is divisible by r,

$$G\left(k\left[\frac{\Omega}{r}\right]\right) = F\left(\left[\frac{k}{r}\right]\Omega\right) \qquad (6.33)$$

If k is not divisible by r, or if r is not an integer, the values of G are different from the values of F but they may be related intuitively by noting that both sets of values are samples of the same continuous function $F^*(\omega)$ defined by (6.9); only the frequency spacing of the samples is different.

Another way to lengthen a sequence artificially is to repeat it. Suppose we create the sequence $h(nT)$ with rN points given by

$$h(nT) = f(((n))T) \qquad \text{for } 0 \le n \le rN - 1 \qquad (6.34)$$

For $h(nT)$ the DFT is given by

$$H\left(k\left[\frac{\Omega}{r}\right]\right) = \sum_{n=0}^{rN-1} h(nT)e^{-j\Omega Tnk/r} = \sum_{n=0}^{N-1} f(nT) \sum_{l=0}^{r-1} e^{-j\Omega T(n+lN)k/r} \qquad (6.35)$$

which may be written

$$H\left(k\left[\frac{\Omega}{r}\right]\right) = \sum_{n=0}^{N-1} f(nT)e^{-j\Omega Tn[k/r]} \left(\sum_{l=0}^{r-1} \nu^l\right) \qquad (6.36)$$

where $\nu = e^{-j(\Omega/r)TNk}$. The sum over l is recognized as the familiar sum of (6.17), so that

$$H\left(k\left[\frac{\Omega}{r}\right]\right) = \begin{cases} rF\left(\left[\frac{k}{r}\right]\Omega\right) & \text{for } k \text{ divisible by } r \\ 0 & \text{otherwise} \end{cases} \qquad (6.37)$$

Table 6.1

$f(nT)$	$\Leftrightarrow$	$F(k\Omega)$
even	$\Leftrightarrow$	even
odd	$\Leftrightarrow$	odd
even and real	$\Leftrightarrow$	even and real
odd and real	$\Leftrightarrow$	odd and imaginary
real	$\Leftrightarrow$	{real part even, imaginary part odd}
imaginary	$\Leftrightarrow$	{real part odd, imaginary part even}
even and imaginary	$\Leftrightarrow$	even and imaginary
odd and imaginary	$\Leftrightarrow$	odd and real

Certain interesting and important theorems relate the DFT of a sequence $f(nT)$ with the DFT of a sequence whose members are the members of $f(nT)$ but permuted. We have already seen the effect of one kind of permutation, rotation, in (6.28). Let us now consider the DFT of the permuted sequence†

$$g(nT) = f(((pn))T) \qquad 0 \le n \le N - 1 \tag{6.38}$$

where p is an integer with no factors in common with N. The DFT of $g(nT)$ is given by

$$G(k\Omega) = \sum_{n=0}^{N-1} f(((pn))T)e^{-j\Omega T nk} \tag{6.39}$$

Now suppose we find the unique integer q (on the range $0 < q < N - 1$) such that‡

$$((pq)) = 1 \tag{6.40}$$

We allow a change of index

$$n = ((qm)) \qquad 0 \le m \le N - 1 \tag{6.41}$$

in (6.39), which gives

$$G(k\Omega) = \sum_{m=0}^{N-1} f(((pqm))T)e^{-j\Omega T qmk} = F(((qk))\Omega) \tag{6.42}$$

Equation (6.42) states that permuting the members of a sequence according to

$$n \to ((pn))$$

before computing the DFT is equivalent to permuting the members of the DFT according to§

$$k \to ((qk))$$

---

† To show that (6.38) is a permutation we must prove a one-to-one correspondence of the integers 0 to $N - 1$ and the sequence $((pn))$, $0 \le n \le N - 1$. Since $0 \le ((pn)) \le N - 1$, it is sufficient to show that $((pn_1)) \ne ((pn_2))$ unless $n_1 = n_2$. If we assume the contrary for some n_1, n_2, then $p(n_1 - n_2)$ must be congruent to zero and thus has a factor N. Since p has none of the factors of N, $n_1 - n_2$ must be a multiple of N. But $|n_2 - n_1| < N$ and so $n_2 = n_1$, which is contrary to the assumption. Thus there is a one-to-one correspondence, and (6.38) is indeed a permutation.
‡ There must be a solution to (6.40) since we proved, in the previous footnote, that $((pn))$ takes on all values 0 to $N - 1$. Thus some value n makes $((pn)) = 1$.
§ It is not always obvious how to solve (6.40) for q. The following modification of Euclid's algorithm was suggested by Brenner. We generate two sequences, r_i and s_i, with $r_0 = N$, $r_1 = p$, $s_0 = 0$, $s_1 = 1$, $r_i = (r_{i-2}) \bmod r_{i-1}$, and $s_i = s_{i-2} + s_{i-1}(r_i - r_{i-2}) - r_{i-1}$. We iterate until $r_i = 0$, and s_{i-1} will be equal to q.

A special case of (6.42) is when $p = q = N - 1$, for which we see that

$$F(((-k))\Omega) = \text{DFT}\{f(((-n))T)\} \tag{6.43}$$

6.3 THE GOERTZEL ALGORITHM

There are many methods for computing a DFT for a single frequency $k\Omega$. The simplest of all is (6.11) itself. This requires N complex multiplications, N complex additions, and the access of N coefficients $e^{-j\Omega Tnk}$. Usually the coefficients would be reconstructed from a stored table of samples of one-quarter period of a sine wave. A complex multiplication requires four real multiplications and two real additions, and a complex addition requires two real additions, so that the total computational effort involved in (6.11) is

$$T_{\text{direct}} = 4N(k_m + k_a + k_e)M \tag{6.44}$$

where k_m is the time required for a real multiplication, k_a is the time required for a real addition, k_e is the time required for other operations in the process, and M is the number of different frequency coefficients to be evaluated.

Another algorithm that computes a single frequency coefficient is the Goertzel algorithm, which is basically a digital-filtering approach. Consider, first, a filter with only a single complex pole at $z = e^{j\Omega Tk}$.

$$H(z) = \frac{1}{1 - e^{j\Omega Tk}z^{-1}} \tag{6.45}$$

This is the first time we have encountered a filter with complex coefficients, but there is no special conceptual difficulty involved, especially since we are filtering complex data. We excite this filter with the sequence $x(nT)$ for which we wish to compute $X(k\Omega)$. The z-transform of $x(nT)$ is $X(z)$ and therefore the output of the filter is

$$y(mT) = \frac{1}{2\pi j} \oint \frac{z^m \sum_{n=0}^{N-1} x(nT)z^{-n}}{z - e^{j\Omega Tk}} \, dz \tag{6.46}$$

and, specifically, when $m = N$, the output is given by the residue at the pole $e^{j\Omega Tk}$.

$$y(NT) = \left[z^N \sum_{n=0}^{N-1} x(nT)z^{-n} \right] \Bigg|_{z = e^{j\Omega T}} \tag{6.47}$$

But since

$$z^N \Bigg|_{z = e^{j\Omega Tk}} = 1$$

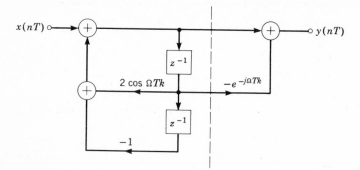

Fig. 6.6 Digital filter to compute one point of DFT by Goertzel algorithm.

we see that

$$y(NT) = X(k\Omega) \tag{6.48}$$

The use of this filter to compute Fourier transforms does not save computations, although it does reduce the coefficient storage required. However if we rewrite (6.45) as

$$H(z) = \frac{1 - e^{-j\Omega Tk}z^{-1}}{1 - 2(\cos \Omega Tk)z^{-1} + z^{-2}} \tag{6.49}$$

we are led to realize the digital filter shown in Fig. 6.6. Since the output of the filter is needed only at N, there is no need to perform the operations indicated by the portion of the network to the right of the dashed line, except at $n = N$. Thus the operations required are N multiplications of a real coefficient by complex data and $2N$ complex additions. The finishing operation is negligible by comparison if the number of data points is large. In terms of real operations,

$$T_{\text{goertzel}} = (2Nk_m + 4Nk_a + Nk'_e)M \tag{6.50}$$

where k'_e is the time required for extra manipulations in the Goertzel algorithm. If, however, the frequencies at which $X(k\Omega)$ is computed can be paired as k, $N - k$, there is an additional savings, since only the zero of the filter in Fig. 6.6 is different for these two frequencies. Thus, if the frequencies can be paired in this way,

$$T_{\text{goertzel}} = N(k_m + 2k_a + k'_e)M \tag{6.51}$$

which is likely to provide a savings of 4 to 1 over the direct computation of (6.11).

For real data, both the direct method and the Goertzel method offer additional savings.

6.4 THE FAST FOURIER TRANSFORM

The direct evaluation of (6.11) requires N^2 complex multiplications and additions, and for moderately large N, say N greater than 1,000, this direct evaluation is rather costly in computer time.   Methods for saving computer time, such as the Goertzel method just considered, have involved very ingenious tricks, mostly based on noticing when a sum of products could be simplified by adding before multiplying or when complex operations could be replaced by real operations.   *These methods have in common that a complete transform of rN points takes r^2 times as long as a transform of N points.*   Thus, even if a method of evaluating (6.11) requires only $\frac{1}{4}N^2$ operations, sequences of more than, say, 10,000 points would require impractical amounts of computation.   However, all such methods have recently been made obsolete by several similar algorithms which we may collectively denote as *fast Fourier transforms* [3, 7, 8, 10, 12, 13].   Although these algorithms do not apply when the number of points in a sequence is a prime, the savings when N is highly composite (has many factors) can be dramatic.†   When N is a power of 2, the fast-Fourier-transform algorithms require a number of computations proportional not to N^2 but to $N \log_2 N$.   Thus, for $N = 1,024$ this is a computational saving of 99 percent, etc.   Efficient application of the algorithms requires that N be highly composite, but for most problems it is possible to impose this restriction on the data to be transformed so that fast-Fourier-transform algorithms may be used.

In the description that follows, we shall use the subscripted notation adopted in (6.27) and Fig. 6.4.   We shall also denote the quantity $e^{-j\Omega T}$ by W.   Thus, (6.11) and (6.12), defining the DFT, become

$$F_k = \sum_{n=0}^{N-1} f_n W^{nk} \tag{6.52}$$

and

$$W = e^{-j(2\pi/N)} \tag{6.53}$$

Equation (6.52) must not be allowed to disguise the fact that W is dependent on N.   For a sequence with $N/2$ points, in (6.52) not only would the upper limit of the sum be $N/2 - 1$, but W would be replaced by W^2.

There are two classes of fast-Fourier-transform algorithms, and each has many modifications.   We shall first consider the method called *decimation in time.*

† In Chap. 7, we shall see how fast Fourier transforms can be computed when the number of points in the sequence is prime.

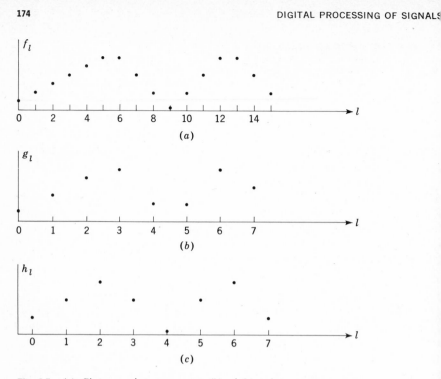

Fig. 6.7 (*a*) Sixteen-point sequence; (*b*) eight-point sequence composed of even-numbered samples; (*c*) eight-point sequence composed of odd-numbered samples.

DECIMATION IN TIME

Suppose the number of samples, N, is divisible by 2. Then it is profitable to consider the discrete Fourier transforms of two shorter sequences: g_l, composed of only the even-numbered samples, and h_l, composed of only the odd-numbered samples. These sequences are illustrated in Fig 6.7 and are given formally as

$$\begin{aligned} g_l &= f_{2l} \\ h_l &= f_{2l+1} \end{aligned} \qquad l = 0, 1, \ldots, N/2 - 1 \qquad (6.54)$$

The DFTs of these sequences are also sequences of $N/2$ points, or, as we have seen, they may be regarded as periodic sequences with period $N/2$ These DFTs may be written

$$\begin{aligned} G_k &= \sum_{l=0}^{N/2-1} g_l(W^2)^{lk} \\ H_k &= \sum_{l=0}^{N/2-1} h_l(W^2)^{lk} \end{aligned} \qquad (6.55)$$

We are really interested in the DFT of the entire sequence, which may, however, be written in terms of g_l and h_l, from (6.54):

$$F_k = \sum_{l=0}^{N/2-1} (g_l W^{2lk} + h_l W^{(2l+1)k}) \qquad (6.56)$$

A simple manipulation of (6.56) yields

$$F_k = \sum_{l=0}^{N/2-1} g_l (W^2)^{lk} + W^k \sum_{l=0}^{N/2-1} h_l (W^2)^{lk} = G_k + W^k H_k \qquad (6.57)$$

The relationship between F_k, G_k, and H_k given by (6.57) has important computational implications. G_k and H_k can, by the direct method, be computed with $(N/2)^2$ operations each, and combining them to give F_k requires an additional N operations, for a total of $N + N^2/2$ operations (an operation here is a complex multiplication and addition). But direct computation of F_k would have required N^2 operations, and so (6.57) reduces the computation by almost a factor of 2 for large N.

In (6.57) the index k runs from 0 to $N-1$. However, G_k and H_k have a period $N/2$ and need be computed only for the range 0 to $N/2 - 1$. Thus for purposes of computation we need a separate "recipe" for F_k for $0 \le k \le N/2 - 1$ and for $N/2 \le k \le N - 1$, namely,

$$F_k = \begin{cases} G_k + W^k H_k & 0 \le k \le N/2 - 1 \\ G_{k-N/2} + W^k H_{k-N/2} & N/2 \le k \le N - 1 \end{cases} \qquad (6.58)$$

This relationship between F_k, G_k, and H_k is pictured in Fig. 6.8, which shows an eight-point DFT reduced to two four-point DFTs.†

It may be that $N/2$ is also divisible by 2, in which case, since G_k and H_k are discrete Fourier transforms, their computation may be reduced in the same way as the computation of F_k was reduced. In fact, the computation of F_k may be successively reduced to that of more and more DFTs with fewer and fewer points with greater and greater total savings. *This is the essence of the fast-Fourier-transform method.* Thus, in Fig. 6.9 the two four-point DFTs of Fig. 6.8 have each been reduced to two two-point DFTs, and in Fig. 6.10 the two-point DFTs have been reduced to one-point DFTs (which are null operations and thus the entire computation of F_k has been reduced to complex multiplications and additions).

† The signal flow graph may be unfamiliar to some readers. Basically it is composed of dots (or nodes) and arrows (transmissions). Each node represents a variable, and the arrows terminating at that node originate at the nodes whose variables contribute to the value of the variable at that node. The contributions are additive, and the weight of each contribution, if other than unity, is indicated by the constant written close to the arrowhead of the transmission. Thus, in Fig. 6.8 the quantity F_7 at the bottom right node is equal to $G_3 + W^7 H_3$. Operations other than addition and constant multiplication must be clearly indicated by symbols other than $\cdot$ or $\longrightarrow$.

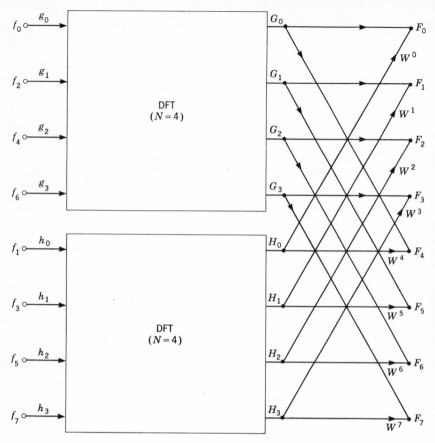

Fig. 6.8 Eight-point DFT reduced to 2 four-point DFTs by decimation in time.

Thus far we have pictured only the case of N a power of 2 or at least divisible by 2. Before proceeding to the case of general N, let us consider the case of N a power of 2 in more detail. An examination of Fig. 6.10 gives considerable information about the practical aspects of computing a discrete Fourier transform. From the figure, it is easy to see that there are 8×3 nodes and $2 \times 8 \times 3$ arrows, corresponding to $N \log_2 N$ additions and $2N \log_2 N$ multiplications. Half of the multiplications can be omitted since the transmissions indicated by the arrows are unity. Half of the remaining multiplications are also easily eliminated, as we shall see in a moment, by utilizing the fact that $W^{N/2} = -1$. Thus, for N a power of 2, $N \log_2 N$ additions and at most $\frac{1}{2}N \log_2 N$ multiplications are required for the computation of the DFT of an N-point sequence. Let us further assume that the input data have been stored in the computer memory in the order

$$f_0, f_4, f_2, f_6, f_1, f_5, f_3, f_7$$

as in Fig. 6.10.   Then the computation of the discrete Fourier transform may be done "in place," that is, by writing all intermediate results over the original data sequence and writing the final answer over the intermediate results.

To see this, suppose each node corresponds to two memory registers (two registers since the quantities are, in general, complex).   The eight nodes farthest to the left then represent the registers containing the shuffled-order input data.   The first step in the computation is to compute the contents of the registers represented by the eight nodes just to the

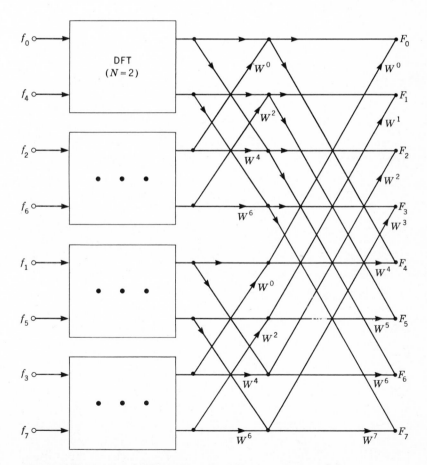

Fig. 6.9 Eight-point DFT further reduced to 4 two-point DFTs.

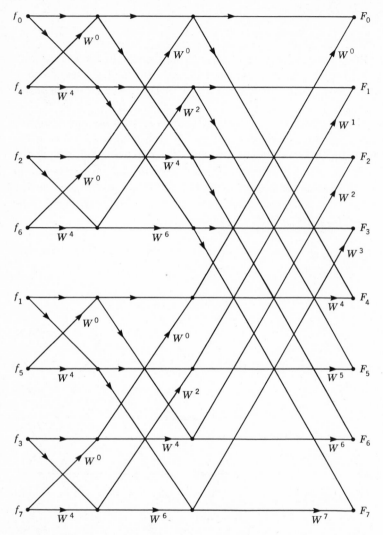

Fig. 6.10 Eight-point DFT completely reduced to complex multiplications and additions by repeated decimation in time. For this diagram inputs are required to be in bit-reversed order.

right of the input nodes. But each pair of input nodes affects only the corresponding pair of nodes immediately to the right, and, if the computation deals with two nodes at a time, the newly computed quantities may be written into the registers from which the input values were taken, since the input values are no longer needed for further computation. The second step, computation of the quantities associated with the next vertical array of nodes to the right, also involves pairs of nodes although these pairs are now two locations apart instead of one. This fact does not change the property of in-place computation, since each such pair of nodes affects only the pair of nodes immediately to the right, so that after a new pair of results is computed it may be saved in the registers that held the old results no longer needed. In the computation for the final array of nodes, corresponding to the values of the discrete Fourier transform, the computation involves pairs of nodes separated by four locations, but the in-place property still holds.

The initial shuffling of the data sequence, f_l, was necessary for the in-place computation. This shuffling is due to the repeated movement of odd-numbered members of a sequence to the end of the sequence during the development of the algorithm. The shuffling has been called *bit reversal* because the sample whose time index is given by the binary number $_2l_1l_0$ must be moved to location $l_0l_1l_2$, and the sample whose time index is $_0l_1l_2$ must be moved to location $l_2l_1l_0$. Note that the initial data shuffling can thus also be done *in place*.

One can also observe from Fig. 6.10 that the two powers of W required by the computation for any pair of nodes differ by 4 or, in the general case, by $N/2$. Since $W^{N/2}$ is equal to -1, it is possible to do the computations for a pair of nodes with only one complex multiplication, one complex addition, and one complex subtraction. Thus, if a subtraction is equivalent in complexity to an addition, our previous count of operations is justified.

The geometry of Fig. 6.10 makes clear the basic idea behind the derivation, replacing the computation of a discrete Fourier transform by two smaller discrete-Fourier-transform computations, and each of these by two still smaller transforms, etc.

Figure 6.10 presupposes a certain data-storage allocation in computer memory before, during, and after the computation. The same equations with different data allocation are described by different flow graphs, and these may seem, at first, to be different versions of the decimation-in-time algorithm. For example, if one imagines that in Fig. 6.10 all the nodes on the same horizontal level as F_1 are interchanged with all the nodes on the same horizontal level as F_4, and all the nodes on the level of F_3 are interchanged with the nodes on the level of F_6, *with the arrows carried along with the nodes*, then a flow graph like that of Fig. 6.11 results. For this rearrangement one need not shuffle the original data into the bit-reversed order, but the resulting spectrum needs to be *unshuffled*. An additional disadvantage might be that the powers of W needed in the computation are in bit-reversed order. Cooley's original description of the algorithm [7] corresponds to the flow graph of Fig. 6.11.

A somewhat more complicated rearrangement of Fig. 6.10 yields the

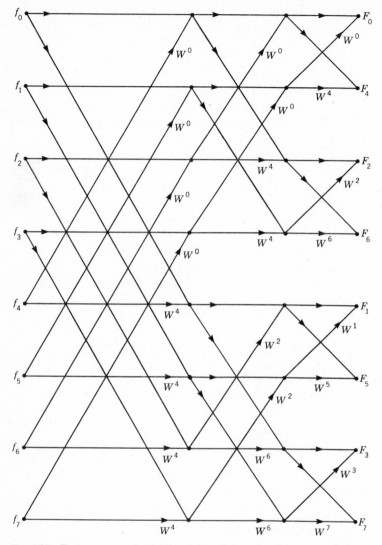

Fig. 6.11 Rearrangement of Fig. 6.10, giving algorithm permitting normally ordered input but bit-reversed output and coefficient order.

signal flow graph of Fig. 6.12. For this case both the input data and the resulting spectrum are in "natural" order, and the coefficients in the computation are also used in a natural order. However, the computation is no longer done in place. Therefore, at least one other array of registers must be provided. This signal flow graph and a procedure corresponding to it are due to Stockham and, independently, to Sande [9].

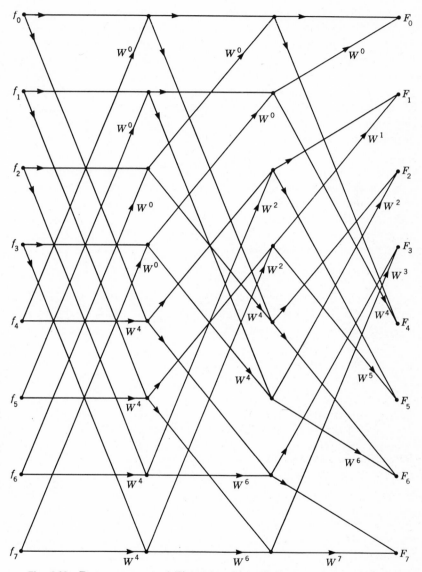

Fig. 6.12 Rearrangement of Fig. 6.10 which eliminates need for bit reversal but which does not permit computation of transform in place.

The three forms of the signal flow graph already given make sense when the data are stored in a random-access memory. Sometimes there are too much data to keep them all in memory at once, and auxiliary storage with sequential-access capability is used. A rearrangement of Fig. 6.10 corresponding to an algorithm of Singleton [13] is given next For this graph, the geometry from stage to stage is the same and calls for sequential access of data with four separate files involved. Thus, four tape units, or a drum with at least four channels, make for practical use of the algorithm as expressed in Fig. 6.13. The algorithm requires that the input be in bit-reversed order but the powers of W are needed in natural order.

Up to this point we have studied in detail only the case of N divisible by 2. However, everything we have done can be generalized to the case of other divisors. Because of the binary organization of digital computers and because it is progressively more difficult to keep track of more and more quantities, algorithms for N not a power of 2 have received less attention than the former case. Nevertheless, if N has a factor p, we can form p different sequences, $g_l^{(i)}$, given by†

$$g_l^{(i)} = f_{pl+i} \qquad (6.59a)$$

each having N/p samples. Each of these sequences has a DFT $G_k^{(i)}$ and the DFT of f_n can be computed from the p simpler DFTs with pN complex multiplications and additions. That is,

$$F_{k+m(N/p)} = \sum_{i=0}^{p-1} G_k^{(i)} W^{i[k+m(N/p)]} \qquad \text{for} \begin{cases} m = 0, 1, \ldots, p-1 \\ k = 0, 1, \ldots, N/p - 1 \end{cases}$$

$$(6.59b)$$

which is identical to (6.58) if $p = 2$. We diagram the case where $p = 3$ $N = 6$ in Fig. 6.14a, and the case where $p = 2$, $N = 6$ in Fig. 6.14b Again, if N/p is composite, the computation of the $G_k^{(i)}$'s can be further simplified.

The collection of methods presented so far have in common the procedure of forming subsequences from the sequence to be transformed each subsequence composed only of every pth point of the original sequence. It is as though the subsequences were formed by sampling the time function at a lower rate or by throwing out $p - 1$ samples of every group of p—thus, the name *decimation in time*.

DECIMATION IN FREQUENCY

Let us now consider another, quite distinct, form of the fast-Fourier transform algorithm, *decimation in frequency*. This form of the algorithm

† The superscript (i) here refers to the ith sequence and is not an exponent.

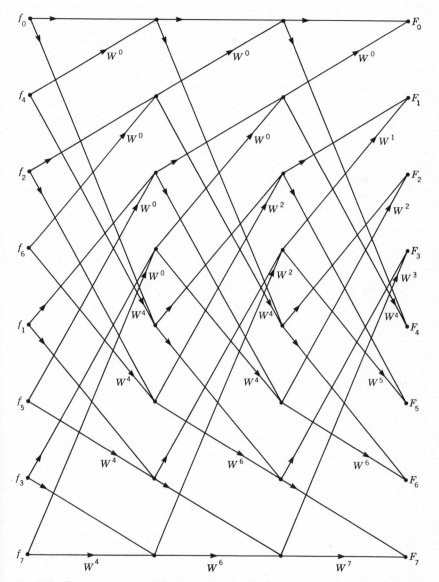

Fig. 6.13   Rearrangement of Fig. 6.10 with identical geometry from stage to stage.

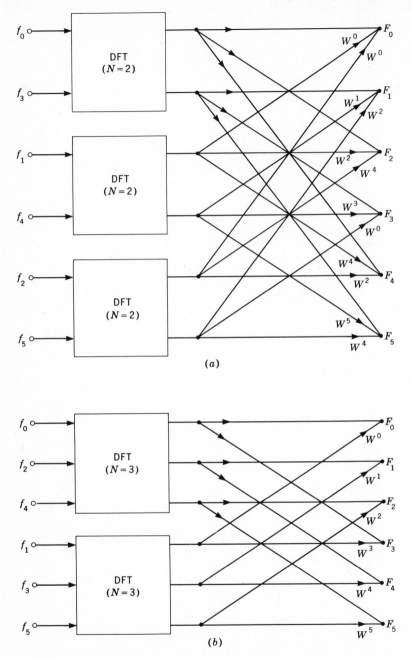

Fig. 6.14 (a) Decimation in time for six-point DFT, with $p = 3$; (b) decimation in time for six-point DFT, with $p = 2$.

was found independently by Sande, Cooley, Stockham, and others. Let the function f_l, with an even number of points, N, be separated into two sequences of $N/2$ points each, say g_l and h_l, where g_l is composed of the first $N/2$ points of f_l and h_l is composed of the last $N/2$ points of f_l. Formally

$$\begin{aligned} g_l &= f_l \\ h_l &= f_{l+N/2} \end{aligned} \qquad \text{for } l = 0, 1, \ldots, N/2 - 1 \qquad (6.60)$$

The N-point DFT, F_k, may now be written in terms of g_l and h_l.

$$F_k = \sum_{l=0}^{N/2-1} (g_l W^{lk} + h_l W^{(l+N/2)k}) \qquad (6.61)$$

or

$$F_k = \sum_{l=0}^{N/2-1} (g_l + e^{-j\pi k} h_l) W^{lk} \qquad (6.62)$$

We now consider the even- and odd-numbered points of F_k separately (therefore, the name decimation in frequency). Replacing k by $2k$ in (6.62) we get

$$F_{2k} = \sum_{l=0}^{N/2-1} (g_l + h_l)(W^2)^{lk} \qquad (6.63)$$

and replacing k by $2k + 1$ in (6.62) we get

$$F_{2k+1} = \sum_{l=0}^{N/2-1} [(g_l - h_l) W^l](W^2)^{lk} \qquad (6.64)$$

which we recognize as the $(N/2)$-point DFTs of the functions $(g_l + h_l)$ and $(g_l - h_l)W^l$. Thus we have found a different way to express the computation of an N-point DFT in terms of two computations of $(N/2)$-point DFTs. In Fig. 6.15 this reduction is diagramed for $N = 8$, and Figs. 6.16 and 6.17 show how successive reductions on the smaller DFTs are carried out as long as the number of points in the subsequences is divisible by 2. Figure 6.17 is a diagram for the computation of a DFT of eight points, completely expressed in terms of complex additions and multiplications. Again the amount of computation is seen to be proportional to $N \log_2 N$. It is interesting to note that Fig. 6.17 has the same geometry as Fig. 6.11, a decimation-in-time version of the algorithm, but the coefficients are different in the two graphs. They occur in natural order in the decimation-in-frequency version. The computation, again, may be done in place. A rearrangement of Fig. 6.17 is given in Fig. 6.18, which has the same geometry as Fig. 6.10 but different coefficients. A more complicated rearrangement yields Fig. 6.19, which, like Fig. 6.12, is a form of the algorithm requiring no bit reversal of input, output, or

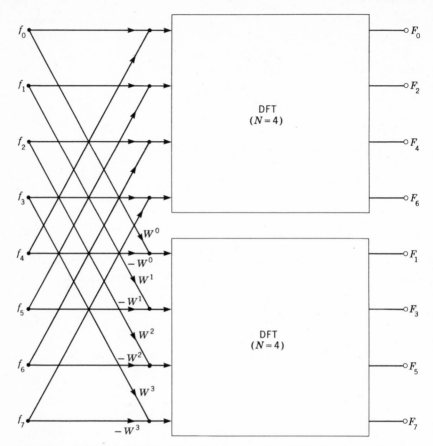

Fig. 6.15 Eight-point DFT reduced to 2 four-point DFTs by decimation in frequency.

coefficients but which cannot be done in place. Finally, Fig. 6.20 is a rearrangement corresponding to an algorithm that is suitable for sequentia access to storage, analogous to Fig. 6.13, but with bit reversal on output samples rather than input samples.

In (6.59a) and (6.59b) we generalized *decimation in time* to include the case where the factors of N are arbitrary. A similar generalization can be made for *decimation in frequency*. The result is

$$F_{pk+r} = H_k^{(r)} \tag{6.65a}$$

where

$$h_l^{(r)} = \sum_{i=0}^{p-1} f_{l+iN/p} W^{(l+iN/p)r} \tag{6.65b}$$

In Fig. 6.21a we diagram (6.65a) and (6.65b) for the case $N = 6$, $p = 2$, and in Fig. 6.21b we diagram the case $N = 6$, $p = 3$.

6.5 THE RELATION BETWEEN DECIMATION IN TIME AND DECIMATION IN FREQUENCY

In Sec. 6.4 we treated *decimation in time* and *decimation in frequency* as distinct, albeit very similar, algorithms. That point of view is useful to the programmer or machine designer, but some additional insight about the nature of the fast-Fourier-transform algorithm may be gained by the different point of view presented in this section. Taking this different point of view does not generally lead to any savings in computational effort but it serves to relate the two forms of the algorithm and suggests a third form in some special circumstances.

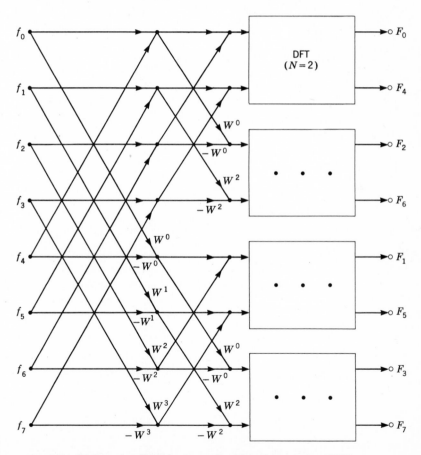

Fig. 6.16 Eight-point DFT further reduced to 4 two-point DFTs.

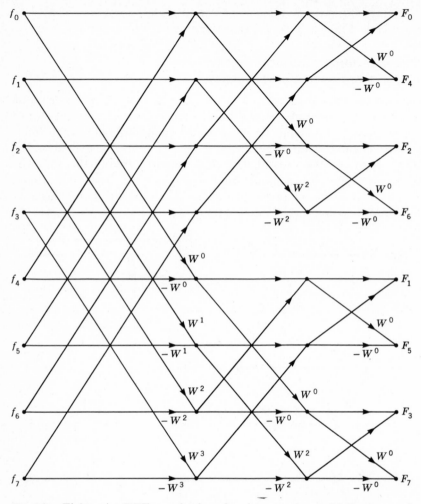

Fig. 6.17 Eight-point DFT completely reduced to complex multiplications and additions by repeated decimation in frequency. For this diagram, inputs are in normal order, as are coefficients, but outputs are in bit-reversed order.

Previously we have thought of the input and output of the DFT as one-dimensional sequences. Suppose the factors of N are p and q. We can write the input sequence as a $p \times q$ two-dimensional array, e.g.,

$$
\begin{array}{cccc}
f_0 & f_1 & \cdots & f_{q-1} \\
f_q & \cdots & \cdots & f_{2q-1} \\
\multicolumn{4}{c}{\cdots\cdots\cdots\cdots\cdots} \\
f_{(p-1)q} & \cdots & \cdots & f_{N-1}
\end{array}
$$

ARRAY 1:

here the general element is f_{l+iq}. The index l is the index within a row, nd the index i is the index within a column. Now we consider *decimation n frequency*. We rewrite Eqs. (6.65) with N/p replaced by q, giving

$$h_l^{(r)} = \sum_{i=0}^{p-1} f_{l+iq} W^{(l+iq)r} \qquad F_{pk+r} = H_k^{(r)} \tag{6.66}$$

, in the first equation of (6.66), we move W^{lr} outside the summation, e can recognize $h_l^{(r)}$ as the *r*th point of the DFT of the *l*th column of array 1.

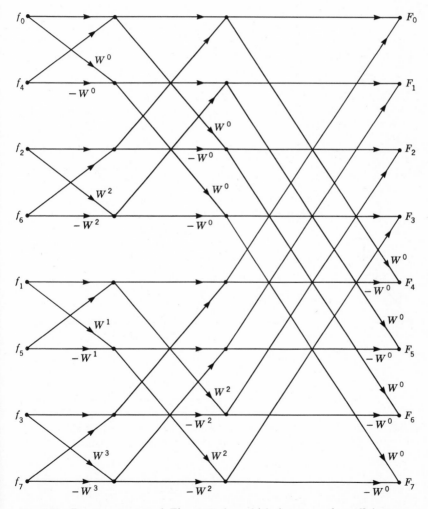

Fig. 6.18 Rearrangement of Fig. 6.17 for which inputs and coefficients are required to be in bit-reversed order but for which outputs are normally ordered.

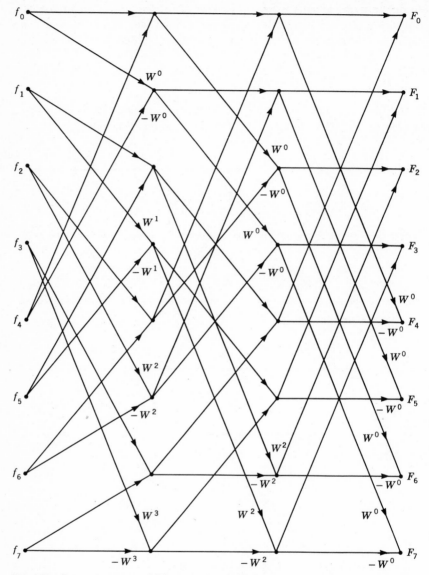

Fig. 6.19 Rearrangement of Fig. 6.17 for which there is no need for bit reversal but for which computation cannot be done in place.

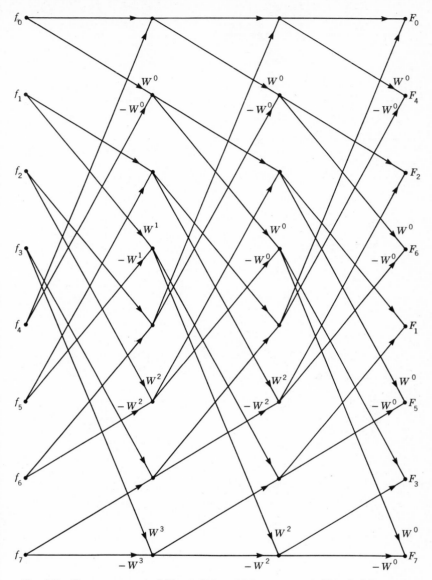

Fig. 6.20   Rearrangement of Fig. 6.17 for which geometry is identical from stage to stage.

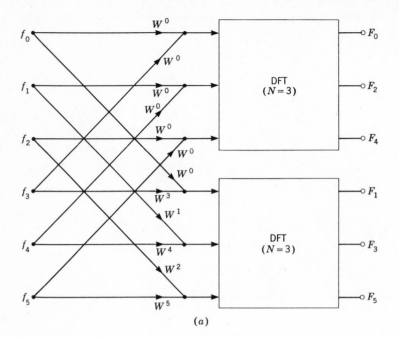

(a)

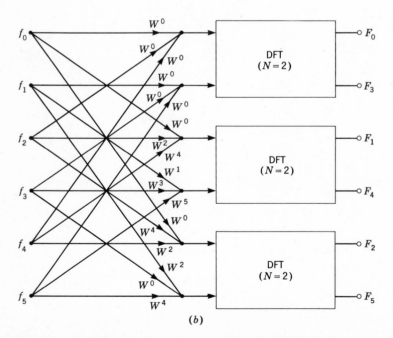

(b)

Fig. 6.21 (a) Decimation in frequency for six-point DFT, with $p = 2$; (b) decimation in frequency for six-point DFT, with $p = 3$.

This is a p-point DFT mapping the *time index* i into the *frequency index* r.

$$h_l^{(r)} = W^{lr} \times \underset{\substack{(p \text{ points} \\ \text{mapping} \\ i \to r)}}{\text{DFT}} \{f_{l+iq}\} \tag{6.67}$$

In all, (6.67) calls for q separate DFTs of p points each. We arrange the results in another array:

$$
\begin{array}{c}
h_0^{(0)} \quad h_1^{(0)} \quad \cdots \quad h_{q-1}^{(0)} \\
h_0^{(1)} \quad \cdots \cdots \cdots \quad \cdots \\
\cdots \cdots \cdots \cdots \cdots \cdots \cdots \\
h_0^{(p-1)} \quad \cdots \cdots \cdots \cdots \quad h_{q-1}^{(p-1)}
\end{array}
$$

ARRAY 2:

The second equation of (6.66) now tells us to compute *the q-point DFTs of each of the p rows of array 2*.

$$F_{pk+r} = \underset{\substack{(q \text{ points} \\ \text{mapping} \\ l \to k)}}{\text{DFT}} \{ W^{lr} \times \underset{\substack{(p \text{ points} \\ \text{mapping} \\ i \to r)}}{\text{DFT}} \{f_{l+iq}\} \} \tag{6.68}$$

Now we think of F_{pk+r} as an array:

$$
\begin{array}{c}
F_0 \quad F_p \quad \cdots \quad F_{(q-1)p} \\
F_1 \quad \cdots \cdots \cdots \quad \cdots \\
\cdots \cdots \cdots \cdots \cdots \cdots \\
F_{p-1} \quad \cdots \cdots \cdots \quad F_{N-1}
\end{array}
$$

ARRAY 3:

where the second DFT mapped the index l into the index k.

To summarize (6.68), then, the input sequence is chopped into several sequences which are arranged as the rows of an array. We then compute a DFT of each of the columns of the array to form the columns of a second array, and we multiply the element in the rth row and the lth column by W^{rl}. Finally, each row of the array is transformed to give the rows of the last array. The columns of the final array may be butted together end to end, giving the N-point DFT. The process is illustrated by Fig. 6.22 for the case of $p = 5$, $q = 3$ and a 15-point DFT.

Now we consider decimation in time, described by

$$F_{k+rq} = \sum_{i=0}^{p-1} G_k^{(i)} W^{i(k+rq)} \qquad g_l^{(i)} = f_{pl+i} \tag{6.69}$$

but let us reverse the meanings of p and q and interchange r with k, and i with l, giving

$$F_{pk+r} = \sum_{l=0}^{q-1} G_r^{(l)} W^{l(r+kp)} \qquad g_i^{(l)} = f_{l+iq} \tag{6.70}$$

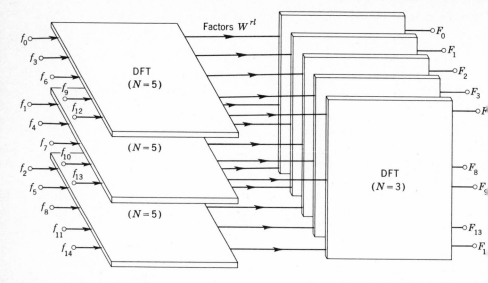

Fig. 6.22 Fast Fourier transform viewed as operation on two-dimensional array.

Equations (6.70) transform f_{l+iq} into F_{qk+r}, as did (6.66), but the process is apparently different. The right-hand equation of (6.70) says to organize array 1 into columns $g_i^{(l)}$. But $G_r^{(l)}$ is the DFT of $g_i^{(l)}$. Therefore, the first step is to form an array whose columns are the DFTs of the columns of array 1. The left-hand equation of (6.70) suggests two steps multiplying the array by W^{rl} and computing the DFTs of each of the rows. Notice, then, that (6.68) is *a description of decimation in time* as well as of *decimation in frequency*. Similarly, Fig. 6.22 is a description of both algorithms. The only difference between the two algorithms is that the factors W^{rl} are incorporated in the second bank of DFTs for *decimation in time*, and they are incorporated in the first bank of DFTs for *decimation in frequency*. Incorporating the factors W^{rl} in one or the other bank of DFTs means that fewer computations are needed for decimation in time or frequency than for a direct implementation of Fig. 6.22. However, one can best appreciate the relationship between the two fast-Fourier-transform methods through Fig. 6.22. The factors W^{rl} have been called *twiddle factors* [9].

In certain special cases, the twiddle factors can be eliminated by tricky permutations on the input and output sequences [4, 10]. This method is based on the circular-shifting theorem together with number theoretical principles. Suppose in array 1 each of the columns is rotated by a different amount, with the lth column rotated by $a \times l$ positions; a is to be determined. The outputs of the first bank of DFTs in Fig. 6.22

will be the same as before except for multiplication by powers of W^q [see (6.28)], i.e.,

$$(W^q)^{-ral} \times \text{DFT } \{f_{l+iq}\}$$

To undo the effects of the input rotations we should change the twiddle factors to

$$W^{rl+qral}$$

Now suppose we are willing to accept, as the output of the second bank of DFTs, not the subsequences $\{F_{pk+r}\}$, but subsequences corresponding to the rows of array 3, with the elements of the rth row rotated by $b \times r$ positions, where b, like a, is to be determined. The circular-shifting theorem has an inverse, which says that this result can be achieved by multiplying the inputs to the second bank of DFTs by

$$(W^{-p})^{brl}$$

and this multiplication can also be accomplished by changing the twiddle factors. When the effects of input and output rotations are incorporated into the twiddle factors, the resulting twiddle factor is

$$W^{rl+qral-prbl}$$

and it is clear that this composite twiddle factor can be eliminated if it is always unity, i.e., if

$$bp - aq = 1 \tag{6.71}$$

If p and q have any common factors, (6.71) cannot be solved for any integers a, b. However, if p and q are mutually prime we can find integers a, b that satisfy (6.70) and, therefore, specify rotations on the input and output that undo the need for twiddle factors. This algorithm is of some use when the number of transform points is a power of 10 and simple algorithms based on a power of 2 and a power of 5 can be effectively combined.

To recapitulate, there are a variety of techniques for reducing the amount of computation required for an N-point discrete Fourier transform. The two basic techniques are decimation in time, in which the transforms of shorter sequences, each composed of every rth sample, are computed and then combined into one big transform, and decimation in frequency, in which short pieces of the sequence are combined in r ways to form r short sequences, whose separate transforms taken together constitute the complete transform. If N has more than two factors, the computation of the transforms required after the reduction can be further simplified by either decimation in time or decimation in frequency. If the input points to a transform are rotated, the output points are multiplied only by certain powers of W, and if the input points to a transform are multiplied by certain powers of W, the output points are only rotated which permits other variations of the basic fast-Fourier-transform algorithms. The flow graphs finally derived may be manipulated in various ways to simplify or at least clarify the problems associated with programming a computer to compute fast Fourier transforms.

Since the inverse discrete Fourier transform is described by an equation essentially identical to (6.11), any method of computing a DFT may easily be modified to compute an inverse DFT. Specifically, W is replaced by W^* (W conjugate), and the result of the computation is multiplied by $1/N$.

6.6 APPLICATION OF THE FAST-FOURIER-TRANSFORM ALGORITHMS

Because of the increased speed with which it is possible to compute DFTs, the number of applications increases also. In this section we shall very briefly mention some of the applications and some of the considerations that occur when the fast Fourier transform is used in these applications.† Some general considerations regarding the programming and use of the fast Fourier transform should be mentioned first. One aspect is accuracy. Although this is still an open question, various workers have reported immense improvements in accuracy‡ when the fast method is compared with the slow method using floating-point arithmetic. These improvements are attributed to the sequencing of computations used by the fast method [5, 9].

The great increase in speed of the fast method relative to all the tricky methods of the past lies in the use of the periodicities of the sine and cosine functions. Previous methods made use only of the symmetries of these functions. It is still possible to introduce savings into the fast-Fourier-transform methods by making use of these symmetries even after the periodicities have been exploited. Thus, multiplications by W^0 and $W^{N/2}$ can be removed from the main loop of a program because they require no actual computer multiplications (which are usually slower than additions). The same is true of multiplications by $W^{N/4}$ and $W^{3N/4}$. Even $W^{N/8}$, and other powers of W at 45° angles to the real and imaginary axes, can be profitably removed from the main program loop because

$$(x + jy) \times (0.707 + j0.707) = 0.707(x - y) + j0.707(x + y)$$

† Unless otherwise indicated, an algorithm limited to 2^n points is assumed.
‡ Measurements of accuracy in fast-Fourier-transform programs often make use of the discrete analog of Parseval's theorem: The sum of the squared magnitudes of the inputs and the sum of the squared magnitudes of the outputs of a DFT are in proportion 1 to N.

requiring only two multiplications rather than four.   Such tricks reduce
the number of multiplications required in the computation by a factor of
approximately $(\log N - 3)/\log N$, which is significant in the case where
$\log N$ is less than about 15.   Some programmers have found that an
algorithm based on powers of 4, 8, and even 16 has much of the simplicity
of a base 2 algorithm but can incorporate these savings in many parts of
the flow graph.   For example, if N is an even power of 2, the base 4
algorithm requires 25 percent fewer complex multiplications than a base
2 algorithm [1].

Another common consideration is the implementation of bit reversal
on a general-purpose computer.   For some applications, as we shall see,
there is no need for bit reversal.   Even when it is necessary, it is always
sufficient merely to count in bit-reversed notation; e.g.,

$$0,\ 8,\ 4,\ 12,\ 2,\ 10,\ 6,\ 14,\ 1,\ 9,\ 5,\ 13,\ 3,\ 11,\ 7,\ 15$$

(for a 16-point transform).   The reversed counter is easier to implement
on a typical computer than the bit reversal of an arbitrary number.
Consider the flow chart of Fig. 6.23.   The most significant bit of X is
tested.   This is the least significant bit of bit-reversed X.   If it is zero,
then it is made 1 and one is finished.   If the two least significant bits of
bit-reversed X are 01, they are made 10 and one is finished, etc.   One-
half of the time only two operations will be necessary, one-quarter of the
time three operations will be needed, one-eighth of the time four operations
will be needed, etc.   The average amount of computation is thus

$$2(\tfrac{1}{2}) + 3(\tfrac{1}{4}) + 4(\tfrac{1}{8}) + 5(\tfrac{1}{16}) + \cdots$$

which converges to three in the limit and is less than three for any finite
number of bits in the counter.   Needless to say, a bit-reversed counter
cannot count past $N - 1$ any more than a conventional counter can.

Among the special applications of the fast Fourier transform, the
most obvious application is power-spectrum estimation.   This has been
discussed in Chap. 4, and we need not elaborate here except to point out
that the basic operations of autocorrelation and Fourier transformation
are simplified by the algorithm [5, 14].   The use of the fast Fourier
transform to compute correlations and convolutions is also extremely
significant and is covered in Chap. 7.   In this application it is necessary
to compute a transform and an inverse transform, and it is possible to use
an algorithm for the transform which produces bit-reversed outputs and
use an algorithm for the inverse transform which accepts bit-reversed
inputs.   In this way it is possible to avoid bit reversing entirely.

Often many transforms are required for a certain result, as in the
processing of continuous, never-ending data.   If the data are real there
is an advantage to computing two transforms at the same time.   Suppose

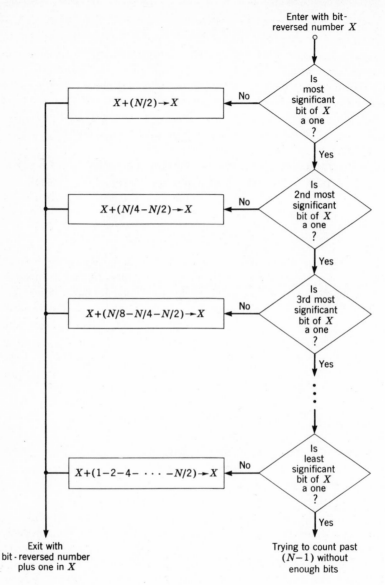

Fig. 6.23   Flow chart for bit-reversed counter.

the real functions are x and y.   A function $z = x + jy$ is formed, with transform $Z = X + jY$.   Then

$$Z = \mathrm{Re}\, X + j\,\mathrm{Im}\, X + j(\mathrm{Re}\, Y + j\,\mathrm{Im}\, Y)$$
$$= (\mathrm{Re}\, X - \mathrm{Im}\, Y) + j(\mathrm{Im}\, X + \mathrm{Re}\, Y)$$

(6.72)

Thus the real part of Z has an even part equal to Re X and an odd part equal to $-$Im Y, and the imaginary part of Z has an even part equal to Re Y and an odd part equal to Im X. But the even and odd parts of a function can be separated with N additions and subtractions, which is a tiny price to pay for getting two transforms for the price of one. In some applications even this price is avoided, as we shall see in Chap. 7. If the problem involves only one real function to be transformed, the even-odd symmetries can still be exploited. Let the even- and odd-numbered points of the original function be the real and imaginary parts of an artificial $(N/2)$-point function which is then transformed. Then the two transforms of the short real functions can be separated as in (6.72) and used as in Fig. 6.8 to compute the complete transform.

In some applications it is not necessary to compute all the points of the transform. Here some additional savings can be made by "pruning" the signal flow graphs we have presented, but the savings tend to be very small.

An interesting application of the fast Fourier transform is in interpolation. The problem is to find the values of points on a curve between the points already known. In the case of an analog band-limited function sampled at $t = nT$, where $1/T$ exceeds the Nyquist rate, the function is given exactly by

$$ f(t) = \sum_{n=-\infty}^{\infty} f(nT) \frac{\sin\left[(\pi/T)(t-nT)\right]}{(\pi/T)(t-nT)} \tag{6.73} $$

This expression calls for an infinite amount of computation for each point on the curve $f(t)$ but, because the $(\sin x)/x$ function falls off rapidly, it is possible to approximate $f(t)$ by some finite sum. Now suppose only N points of the function are known. It is impossible for a function that is zero outside the range 0 to NT to be a band-limited function. Hence precise band-limited interpolation is impossible for this case. Let us, therefore, assume that the known N points represent one period of a periodic band-limited function. Then we may estimate the function at r times as many points (where r need not be an integer) by computing the N-point DFT and inserting $(r-1)N$ zeros in the middle of the sequence $F(k\Omega)$. The inverse transform will then have rN points corresponding to a band-limited interpolation of the periodic function. This is illustrated in Fig. 6.24.

6.7 SUMMARY

We have defined the DFT and developed its properties in analogy with those of the continuous Fourier transform. These are exact properties,

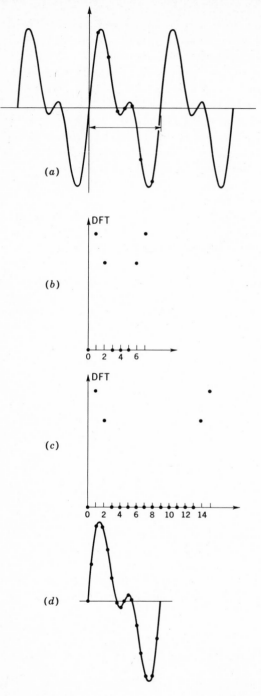

(a)

(b)

(c)

(d)

Fig. 6.24 Band-limited interpolation by DFT: (a) sampled function; (b) its DFT; (c) modified DFT; (d) interpolated function.

not approximations to those of the continuous Fourier transform. After examining two conventional algorithms for computing DFTs, we introduced the fast Fourier transform, which requires a computation time far less than the conventional methods. For the number of data points a power of 2, there are two basic algorithms, decimation in time and decimation in frequency. Each algorithm has many variations, depending on the allocation of data storage before, during, and after the computation. Algorithms for computation of a spectrum in the same memory as the data require a permutation of input samples, output samples, or required constants, but the elimination of these permutations requires extra memory for the computation.

The fast Fourier transform has made computation of discrete Fourier transforms practical; this has, in turn, given practical significance to several of the properties of the DFT. At the time of this writing it appears that the possible applications of discrete Fourier transforms have probably not all been discovered. However, the digital-signal-processing field has already been changed substantially by the availability of the algorithm, as we shall see in the next two chapters.

REFERENCES

1. Bergland, G. D.: A Fast Fourier Transform Algorithm Using Base Eight Iterations, *Math. Comput.*, **22**: 275–279 (April, 1968).
2. Brenner, N. M.: Three Fortran Programs That Perform the Cooley-Tukey Fourier Transform, *Mass. Inst. Technol., Lincoln Lab. TN*-1967-2, July 28, 1967.
3. Cochran, W. T., et al.: What Is the Fast Fourier Transform?,† *IEEE Trans. Audio*, **15**(2): 45–55 (June, 1967).
4. Cooley, J. W., P. Lewis, and P. Welch: Historical Notes on the Fast Fourier Transform,† *IEEE Trans. Audio*, **15**(2): 76–79 (June, 1967).
5. Cooley, J. W., P. Lewis, and P. Welch: The Fast Fourier Transform and Its Applications, *IBM Res. Paper RC*-1743, Feb. 9, 1967.
6. Cooley, J. W., P. Lewis, and P. Welch: Application of the Fast Fourier Transform to Computation of Fourier Integrals, Fourier Series, and Convolution Integrals, *IEEE Trans. Audio*, **15**(2): 79–84 (June, 1967).
7. Cooley, J. W., and J. W. Tukey: An Algorithm for the Machine Computation of Complex Fourier Series, *Math. Comput.*, **19**: 297–301 (April, 1965).
8. Danielson, G. C., and C. Lanczos: Some Improvements in Practical Fourier Analysis, and Their Application to X-ray Scattering from Liquids, *J. Franklin Inst.*, **233**: 365–380, 435–452 (April, May, 1942).
9. Gentleman, W. M., and G. Sande: Fast Fourier Transforms—For Fun and Profit, presented at 1966 Fall Joint Computer Conference, *AFIPS Proc.*, **29**: 563–578 (1966).
10. Good, I. J.: The Interaction Algorithm and Practical Fourier Series, *J. Roy. Statist. Soc., Ser. B.*, **20**: 361–372 (1958); Addendum, **22**: 372–375 (1960).

† References 3 and 4 were reprinted in *Proc. IEEE*, **55**(10): (October, 1967).

11. Helms, H. D.: Fast Fourier Transform Method of Computing Difference Equations and Simulating Filters, *IEEE Trans. Audio,* **15**(2): 85–90 (June, 1967).
12. Rudnick, P.: Note on the Calculation of Fourier Series, *Math. Comput.,* **20**: 429–430 (July, 1966).
13. Singleton, R. C.: A Method for Computing the Fast Fourier Transform with Auxiliary Memory and Limited High Speed Storage, *IEEE Trans. Audio,* **15**(2): 91–98 (June, 1967).
14. Stockham, T. G., Jr.: High Speed Convolution and Correlation, presented at 1966 Spring Joint Computer Conference, *AFIPS Proc.,* **28**: 229–233 (1966).

7
High-speed Convolution
and Correlation with
Applications to Digital Filtering

by Thomas G. Stockham, Jr.

7.1 INTRODUCTION

The discussions of this chapter are intended to focus attention upon a new method for summing lagged products which effects savings in computing time [5, 6, 8, 14]. These savings can be very large indeed since the speedup factor for lags involving N products is proportional to $N/\log N$. The requirement for summing lagged products arises frequently in automatic computation, sometimes under a different name, a fact that broadens the significance of the method. Besides obvious economic implications, the speedup advantage makes interactive programming applicable to a wider class of problems and provides an opportunity for explorations previously deemed too ambitious.

The summation of lagged products usually arises in problems involving convolution or correlation. Such problems are very common in electrical engineering, statistics, physics, and optics. Some examples are signal filtering, spectral estimation, image formation, and problems involving field theory such as acoustics, diffusion, electricity, and magnetism. In this chapter we are concerned primarily with the first two.

The methods of this chapter are based upon the use of the fast-Fourier-transform algorithm (FFT) discussed in Chap. 6. Without the FFT they would not exist. At the time of their discovery, reasoning was something like the following.

Since multiplication of Fourier transforms corresponds to convolution of the transformed functions, the process of convolution can be exchanged for three Fourier transformations and a multiplication. If the transforms can be performed more rapidly than the convolution itself, the exchange has a practical value as well as a theoretical one. Since an FFT takes time proportional to $N \log N$ and a convolution takes time proportional to N^2, it is not difficult to see that savings always ensue for sufficiently large N and grow rapidly thereafter.†

Convolution and correlation are related by a very simple change of variables which merely turns one function backward with respect to its independent variable. Thus the discussions above apply equally well to correlation.

Subsequent efforts to translate this reasoning into a set of specific algorithms were confronted with a most important obstacle that is not obvious from the outset. This obstacle is manifested by the fact that the FFT applies to the discrete Fourier transform (DFT) only and that multiplication of DFTs corresponds to convolution of periodic functions, not aperiodic ones. A convolution of periodic functions can be alternatively regarded as a circular convolution in which values of the kernel that are shifted from one end of a period are circulated into the other end, thus introducing what might be termed interperiod interference.

This obstacle is overcome by finding a way to force a periodic convolution to yield results numerically identical to those of an aperiodic one. This is done by augmenting each cycle of both functions with zero values to provide a kind of "grace band" during which the interperiod interference assumes zero numerical value. This grace band can be made large or small, depending upon how much of the periodic convolution is to be rendered into an *aperiodic equivalent*.

With these issues in mind, it is simple to follow the detailed steps out of which high-speed convolution and high-speed correlation are made. In fact the reader will undoubtedly see some alternative steps which present themselves but which are not brought out explicitly for the sake of clarity.

7.2 THE PRODUCT OF DFTs: PERIODIC CONVOLUTION

We have seen in Sec. 6.2 that the product of two DFTs,

$$V(k\Omega) = X(k\Omega) Y(k\Omega) \tag{7.1}$$

† One empirical measurement [14] exhibited the crossover for values of N near 28.

; the DFT of a sequence that was called the circular or periodic convolu-
ion of $x(nT)$ and $y(nT)$.

$$v(mT) = \sum_{n=0}^{N-1} x(nT)y(((m-n))T) = \sum_{n=0}^{N-1} x(((m-n))T)y(nT)$$
(7.2)

he time required to compute $v(mT)$ from either equation in (7.2) is

$$T_{\text{circ}} = k_{\text{circ}}N^2$$
(7.3)

where k_{circ} is a constant of proportionality related to the time necessary to
nd two samples, multiply them together, and add them to a partial sum.
However, we recall that the DFTs of $x(nT)$ and $y(nT)$ and also the inverse
DFT of the product can be computed in times that are proportional to
$N \log N$, that is,

$$T_{fft} = k_{fft}N \log_2 N$$
(7.4)

where k_{fft} is the constant of proportionality. Therefore, if one computes
$X(k\Omega)$ and $Y(k\Omega)$, using the FFT, multiplies them together to get $V(k\Omega)$,
and then evaluates $v(mT)$ using the inverse FFT, the total time required
s

$$T_{fp} = 3k_{fft}N \log_2 N + k_m N = 3k_{fft}N(\log_2 N + \lambda_m)$$
(7.5)

where $k_m N$ is the time required to compute (7.1) and $\lambda_m = k_m/3k_{fft}$.
This assumes N is a power of 2. Similar savings would be possible
provided N is a highly composite number.†

It is sometimes desirable to compute circular convolutions in
applications. However, in most applications, the values of the kernel
hat are shifted from one end of the summation interval are not expected
o be shifted in from the other end; in most applications *aperiodic* con-
olution or correlation is desired.

.3 APERIODIC CONVOLUTION

When suitable steps are taken, a periodic convolution can be used to
ompute an aperiodic convolution provided that each aperiodic sequence
has zero amplitude outside some single finite aperture.‡ A pair of
periodic functions, α_n and β_n, satisfying these conditions is shown in
Fig. 7.1. Also shown is γ_n, the result of the aperiodic convolution
of α_n and β_n. γ_n can be obtained entirely, or in part, from the result

---

The reader should not be led to believe that a periodic convolution cannot be com-
uted by using high-speed convolution techniques when the period N is not composite.
t can; however, not directly as described thus far.
‡ There may, of course, be any number of zero values inside the aperture.

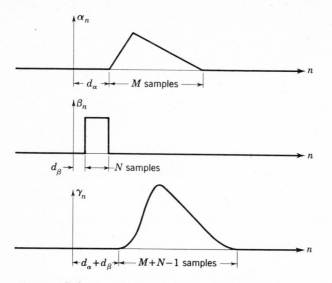

Fig. 7.1 Pair of aperiodic sequences of finite aperture, α_n and β_n, and their aperiodic convolution α_n.

of convolving suitable periodic sequences derived from α_n and β_n, that is, from a circular convolution of suitably formed sequences $\hat{\alpha}_n$ and $\hat{\beta}_n$, shown in Fig. 7.2.   These derived sequences are formed by ignoring the offset delays d_α and d_β and repeating the aperiodic functions α_n and β_n with a period L selected large enough to avoid any overlap.

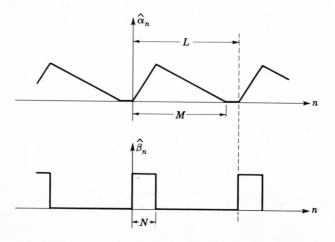

Fig. 7.2   Periodic sequences $\hat{\alpha}_n$ and $\hat{\beta}_n$ derived from α_n and β_n.

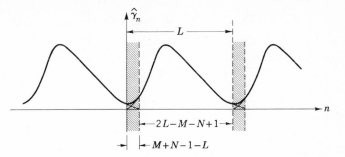

Fig. 7.3 Periodic convolution of $\hat{\alpha}_n$ and $\hat{\beta}_n$, including interfering and noninterfering intervals.

Figure 7.3 shows $\hat{\gamma}_n$, the periodic convolution of $\hat{\alpha}_n$ and $\hat{\beta}_n$. $\hat{\gamma}_n$ can also be formed by ignoring the offset delay of d_γ, repeating it with period L, and summing values within any interfering intervals. Outside these interfering intervals $\hat{\gamma}_n$ is an exact replica of the desired γ_n. If L is made large enough so that $L \geq M + N - 1$, there are no interfering intervals. Thus we see that γ_n may be obtained entirely or in part by extracting the period of $\hat{\gamma}_n$ starting at $n = 0$, subjecting it to the offset delay $d_\alpha + d_\beta$, and disregarding any portion corresponding to an interfering interval.

In order to use the FFT to compute $\hat{\gamma}_n$, L must be adjusted to a power of 2. There are at most only two choices for L that make sense. If $M \geq N$, as in Figs. 7.2 and 7.3, the smallest allowable L value will be 2^v such that

$$2^{v-1} < M \leq 2^v \tag{7.6}$$

In the event that it is also true that

$$M + N - 1 \leq 2^v \tag{7.7}$$

this smallest value automatically avoids interfering intervals. If, however,

$$M + N - 1 > 2^v \tag{7.8}$$

then interfering intervals will result unless L is selected to be 2^{v+1}. Larger L values are never required. In either case

$$M \leq L \leq 2(M + N - 2) \tag{7.9}$$

and the time T_{fa} required to compute a fast aperiodic convolution is bounded by

$$T_{fa} \geq 3k_{fft}M(\log_2 M + \lambda_m)$$
$$T_{fa} \leq 6k_{fft}(M + N - 2)[\log_2(M + N - 2) + \lambda_m + 1] \tag{7.10}$$

regardless of what values may be assumed by M and N.

7.4 PARTIAL RESULTS AND SECTIONED RESULTS

Often, especially in cases when M is considerably larger than N, interest is limited to a relatively small contiguous segment of γ_n. If this is the case, there is no point in computing all of γ_n only to throw away much of it.†

Let us assume that we are interested in γ_n on a segment containing S samples starting at the point $n = n_0$. We may obtain them more efficiently by selecting a segment of α_n containing $S + N - 1$ samples starting at the point $n = n_0 - N + 1$ and using it to form an abbreviated α_n with a new smaller M such that

$$M = S + N - 1 \tag{7.11}$$

The $\hat{\gamma}_n$ thus produced contains the partial results that are desired. They are $\hat{\gamma}_{N-1}$ to $\hat{\gamma}_{S+N-2}$ inclusive, corresponding numerically to γ_{n_0} to γ_{n_0+S-1} inclusive. In this situation there is only one suitable value for L such that

$$S + N - 1 \leq L \leq 2(S + N - 2) \tag{7.12}$$

Thus the time T_{fap} required to compute a fast partial aperiodic convolution is

$$T_{fap} \geq 3k_{fft}(S + N - 1)[\log_2 (S + N - 1) + \lambda_m]$$
$$T_{fap} \leq 6k_{fft}(S + N - 2)[\log_2 (S + N - 2) + \lambda_m + 1] \tag{7.13}$$

An extremely common application of the process of convolution is that of signal filtering. In this application it is usual for M to be very large, perhaps indefinitely large, since α_n represents a signal such as speech, seismic tremors, or perhaps even television pictures of Mars. In a permanent installation where a signal must be filtered as it is generated, M must be considered to be indefinitely large. If α_n were produced at the rate of 1,000 samples per second, one year of continuous operation would result in $M \approx 10^{10}$ samples.

An attempt to use the methods described thus far to effect the desired signal filtering would be hampered by two major obstacles. The first is that no computer memory would be large enough to handle the $\hat{\alpha}_n$ and $\hat{\beta}_n$ involved. The second is that results would be subject to an enormous delay since the last sample of α_n would have to be known before any filtered results could be produced.

Both the difficulties mentioned above can be relieved by a scheme known as sectioning. An additional benefit that arises as a by-product of sectioning is an additional saving in time over that indicated in (7.10).

† A common example of this situation occurs when an autocorrelation function is to be computed for use in power-spectrum estimation.

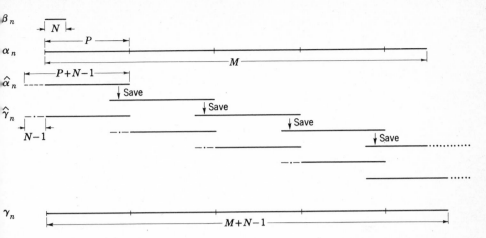

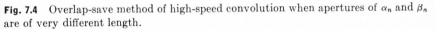

Fig. 7.4 Overlap-save method of high-speed convolution when apertures of α_n and β_n are of very different length.

Sectioning can be performed in two ways. The first method, based upon the principle of partial results discussed above, is called the *overlap-save* method. The second method is called the *overlap-add* method [14].

In the overlap-save method one imagines the result γ_n to be sectioned into parts, each of which contains P samples so that $P + N - 1 = L$, a power of 2. This requires K complete parts, where

$$K = \text{largest integer} \leq \frac{M + N - 1}{P} \qquad (7.14)$$

and most probably one incomplete part of length P_r, where

$$P_r = M + N - 1 - PK \qquad (7.15)$$

Each part of γ_n is formed in turn via the methods of partial results.† The time T_{sect} required is

$$\begin{aligned} T_{\text{sect}} \leq\ & [1 + 2(K + 1)]k_{fft}(P + N - 1)[\log_2 (P + N - 1)] \\ & + (K + 1)k_{\text{aux}}(P + N - 1) \\ \approx\ & (2K + 3)k_{fft}(P + N - 1)[\log_2 (P + N - 1) + \lambda_{\text{aux}}] \end{aligned}$$
$$(7.16)$$

where $k_{\text{aux}} (P + N - 1)$ is the time required to form an abbreviated $\hat{\alpha}_n$ from α_n, to compute (7.1), and to extract a part of γ_n from $\hat{\gamma}_n$. λ_{aux} is $k_{\text{aux}}/2k_{fft}$.

The overlap-save method derives its name from the fact that each

† Note that the DFT of $\hat{\beta}_n$, assumed the shorter sequence, need be computed only once and applied to each of the sections of α_n.

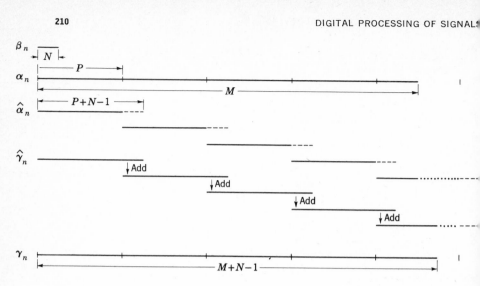

Fig. 7.5 Overlap-add method of high-speed convolution when apertures of α_n and β_n are of very different length.

abbreviated $\hat{\alpha}_n$ represents a section of α_n that overlaps its neighbor by $N - 1$ samples. Each new $\hat{\alpha}_n$ can be formed by combining P new input samples with $N - 1$ old input samples saved from the overlap. This situation is depicted in Fig. 7.4. Notice that the saved samples are at the end of old $\hat{\alpha}_n$'s and at the beginning of new ones. The $N - 1$ samples at the beginning of each $\hat{\gamma}_n$ are meaningless since they correspond to interfering intervals. They are discarded during the formation of γ_n.

The overlap-add method of performing a sectioned convolution is substantially the same as the overlap-save method except that α_n is sectioned and γ_n is overlapped. The sections of α_n again contain P samples, and L is unchanged. This time, however,

$$K = \text{largest integer} \leq \frac{M}{P} \tag{7.17}$$

and

$$P_r = M - PK \tag{7.18}$$

which are changes of minor consequence.

Each $\hat{\alpha}$ is formed by selecting the corresponding section of α and adding $N - 1$ zero-valued samples at the end. Each resulting $\hat{\gamma}_n$ has no interfering interval and represents a complete aperiodic convolution of β_n with a section of α_n. Therefore, each $\hat{\gamma}_n$ must be additively overlapped with its neighbors to form the desired γ_n. This situation is depicted in Fig. 7.5.

The time required to perform the overlap-add method is substantially the same as for the overlap-save method. There is something to be gained f K turns out smaller for the overlap-add method but this may be lost because it requires a small additional time to add overlaps rather than save them.

In the event that P_r is a very small number, it may be attractive to process the last section either with a reduced value of L or by using an ordinary convolution algorithm. If $M \gg N$, it is a small loss to forget this kind of saving even if $P_r = 1$.

The precise value of λ_{aux} depends upon what forms of storage are used to handle the sections of α_n and γ_n as they are needed and produced. Surprisingly enough, λ_{aux} will always be small compared with $\log_2$ $(P + N - 1)$ as long as the auxiliary storage is of tape, disk, or drum speed. A figure of 1.5 to 2.0 for λ_{aux} is typical [6, 13, 14].

7.5 PAIRING IN REAL CONVOLUTION

Most signal-filtering operations involve a real signal α_n and a real impulse response β_n. In the above derivations no such restriction applies to α_n and β_n, and they may be complex functions. In other words, γ_n is really four real convolutions combined two at a time. More accurately,

$$\gamma_n = \text{Re } \gamma_n + j \text{ Im } \gamma_n = \text{Re } \alpha_n * \text{Re } \beta_n - \text{Im } \alpha_n * \text{Im } \beta_n$$
$$+ j(\text{Im } \alpha_n * \text{Re } \beta_n + \text{Re } \alpha_n * \text{Im } \beta_n) \quad (7.19)$$

If $\text{Im } \beta_n = 0$ the impulse response is real, and (7.19) reduces to two separate real convolutions involving β_n. Thus for real signals and impulse responses the methods derived thus far are capable of either double duty or doubled performance. Actually there are three possibilities. Two signals may be processed by the same impulse response, two impulse responses may process the same signal, or one impulse response may process one signal in half the time.

Of these three possibilities the last is probably the most important. It is implemented by combining adjacent sections of $\hat{\alpha}_n$'s into complex composites. This is accomplished by letting odd-numbered sections of $\hat{\alpha}_n$ be used as real parts and even-numbered sections of $\hat{\alpha}_n$ as imaginary parts. The convolution of these complex composite $\hat{\alpha}_n$'s with the real $\hat{\beta}_n$ produces complex composite $\hat{\gamma}_n$'s, the real parts of which are the odd-numbered sections of $\hat{\gamma}_n$ and the imaginary parts of which are the even-numbered sections of $\hat{\gamma}_n$.

Most precisely this device reduces the time required by changing (7.14) to

$$K = \text{largest integer} \leq \frac{M + N - 1}{2P} \quad (7.20)$$

and (7.17) to

$$K = \text{largest integer} \leq \frac{M}{2P} \qquad (7.21$$

Although (7.16) remains unchanged, K has been reduced.

7.6 CHOOSING THE SECTION LENGTH

For a specific pair of values M and N, the section length P should b
chosen to minimize T_{sect}. Since $P + N - 1$ must be a power of 2, on
may evaluate (7.16) for the few values of P that are compatible with thi
constraint and the size of available memory and then select the optimun
length [5, 8, 14].

7.7 EFFECTS OF LARGE N

N, the number of samples in the impulse response, is limited by the
amount of high-speed memory available. Experiments have shown [14
that N must be limited to occupy about one-eighth of the memory not use
for the program. If N is larger than this limit, the impulse response β
must be split into packets, each of which is used as a separate impulse
response. The results corresponding to all packets are then added
together after each has been shifted by a suitable number of packet
widths. This additive process can be accomplished very easily by using
auxiliary storage.

For very large numbers of sections, K, (7.16) can be modified to
involve M explicitly instead of implicitly through K. This may be done
by using the real convolution value of K given in (7.21):

$$T_{\text{fast}} \approx k_{fft}M \frac{P + N - 1}{P} [\log_2 (P + N - 1) + \lambda_{\text{aux}}] \qquad (7.22$$

Paying no attention to the optimization of P as described above, we may
conservatively select $P \approx N$ and through the use of $\lambda_{\text{aux}} = 2$ develop

$$T_{\text{fast}} \approx 2k_{fft}M(\log_2 N + 3) \qquad (7.23$$

which provides a very simple formula for estimating signal-filtering time
in terms of k_{fft}, N, and M.

Perhaps slightly more interesting is the formula for the filtering
period in which the factor M has been divided out of (7.23).

$$\tau_{\text{fast}} \approx 2k_{fft}(\log_2 N + 3) \qquad (7.24$$

where τ_{fast} represents the time required to filter a single sample of α_n
Thus $(\tau_{\text{fast}})^{-1}$ represents the frequency in samples per second at which the
filter operates.

The IBM 7094 model II computer processing floating-point data can asily realize a value of 100 μsec for k_{fft}, and with some specialized coding 1 machine language a reduction to 75 μsec has been achieved. Using the maller figure and $N = 8,192$, we obtain $\tau_{fast} = 2.4$ msec, which corresponds to a filtering rate of over 400 samples per second. This value of J is about the largest that can be managed on a machine with 64,000 ords of high-speed memory and provides an extremely complex filtering apability. For a real-time application the impulse response would be nore than 19 sec long.

Tests have shown [14] that standard methods for computing convolutions require about the same time as the high-speed methods when N rops to just below 32. For this value of N, $\tau_{fast} = 1.2$ msec, only half f the value for $N = 8,192$. This weak dependence upon N is an xtremely important characteristic of the high-speed convolution method.

Within the next few years the array computer will make an enormous ifference in the speed with which an FFT can be computed. By the arly 1970s there is great reason to expect a speedup of more than three rders of magnitude from the figures quoted above. With increased nemory sizes coming along simultaneously, real-time digital filtering at negahertz bandwidths employing frequency characteristics of very great omplexity will be possible.

8 DISCRETE FOURIER TRANSFORMS EXPRESSED S CONVOLUTIONS OR CORRELATIONS

Ve have already seen that there is an intimate connection between DFTs nd lagged products, either circular or periodic, and that this connection ermits computation of large numbers of lagged products in a time proortional to $N \log N$. In this section we shall expose a connection that ermits the DFT to be computed as a lagged product; this has applications) situations where the FFT is not directly applicable to the computation f a spectrum. There are, in fact, two distinctly different relations, the rst being quite general and the second being less general but no less nteresting. We shall consider first a relation suggested by the continuous echnique of *chirp filtering* [4].

Beginning with the definition of a DFT,

$$F_k = \sum_{n=0}^{N-1} f_n W^{nk} \tag{7.25}$$

e can multiply the right side by the quantity $W^{\frac{1}{2}(n^2+k^2)}$ and its eciprocal, leaving the equation unchanged. But we can gather and regroup factors in (7.25) to give

$$F_k = W^{-\frac{1}{2}k^2} \left[\sum_{n=0}^{N-1} (f_n W^{-\frac{1}{2}n^2}) W^{\frac{1}{2}n^2} W^{nk} W^{\frac{1}{2}k^2} \right] \qquad (7.26)$$

and since

$$W^{\frac{1}{2}n^2} W^{nk} W^{\frac{1}{2}k^2} = W^{\frac{1}{2}(n+k)^2} \qquad (7.27)$$

we get the following expression for a DFT:

$$F_k = W^{-\frac{1}{2}k^2} \left[\sum_{n=0}^{N-1} (f_n W^{-\frac{1}{2}n^2}) W^{\frac{1}{2}(n+k)^2} \right] \qquad (7.28)$$

Equation (7.28) suggests a three-step process for computing F_k:

1. Multiply f_n by $W^{-\frac{1}{2}n^2}$ to give a new sequence ξ_n.
2. Compute ζ_k, the correlation for lag k of the sequences ξ_n and $W^{\frac{1}{2}n^2}$.
3. Multiply ζ_k by $W^{-\frac{1}{2}k^2}$ to give the DFT, F_k.

The major step, computing the correlation,† can be implemented by high-speed techniques, so that we have arrived at the seemingly futile result that we can use an FFT to compute a DFT in a somewhat roundabout manner. However, this result is not really useless, because we have bypassed several of the restrictions inherent in the computation of a DFT directly by FFT techniques. For example, there is no restriction that the number of points in f_n be highly composite since an FFT based on a power of 2 can be used to compute lagged products with any number of points. This means that a DFT for a prime number of points can be computed in a time proportional to $N \log N$ rather than N^2, as was formerly thought to be required, although the constant of proportionality is several times greater than that associated with the FFT. But this is not all. The derivation of (7.28) did not in any way depend on the nature of W. Therefore, whereas we were formerly restricted, by the FFT, to computing the z-transform of a sequence at points spaced $2\pi/N$ along the unit circle, we can now compute the z-transform at any equal spacing along any

---

† Since $W = e^{-j2\pi/N}$, it has been shown [4] that, for N a perfect square, say M^2, there is a recursive filter whose impulse response $W^{-\frac{1}{2}n^2}$ can be convolved with ξ_n to perform the required correlation. This may be seen from the z-transform

$$H(z) = \sum_{n=0}^{\infty} e^{+j\pi n^2/M^2} z^{-n} = \sum_{m=0}^{M-1} z^{-m} e^{j\pi m^2/N} \frac{1}{1 + e^{j2\pi m/M} z^{-M}}$$

which represents a bank of M recursive filters with delays and weights. This leads to an interesting DFT algorithm requiring a time proportional to $N^{3/2}$ for N a perfect square.

portion of the unit circle and still retain the $N \log N$ dependence in the computation time. In fact, we can even compute the z-transform at points on a spiral

$$z = W^k \qquad k = k_0, k_0 + 1, k_0 + 2, \ldots, k_0 + N - 1$$

by making the magnitude of W be different from unity. If W is real, this spiral becomes a set of exponentially spaced points on the real axis.

We have now seen that we can achieve the FFT kind of efficiency in computing the DFT of a function when the number of points is prime. The second technique alluded to in the introductory paragraph of this section is applicable *only* when the number of points is prime (although it may be modified to include the case of any power of any odd prime). Whereas (7.28) was suggested by chirp filtering, the method now to be revealed seems to be entirely distinct from any known relationship in continuous-Fourier-transform theory. We begin with a discussion of a property of certain integers, including primes: the existence of primitive roots.

A primitive root is defined as follows: Let the number of integers less than N and mutually prime to N be φ. Then if the sequence

$$\{((g^1)), ((g^2)), ((g^3)), \ldots, ((g^\varphi))\}$$

contains all those φ numbers (and of course neither any others nor any repetitions) g is said to be a primitive root of N. If N is prime, $\varphi = N - 1$, and this definition reduces to a statement that the sequence

$$\{((g^n))\} \qquad n = 1, 2, \ldots, N - 1$$

is a permutation of the integers $1, 2, \ldots, N - 1$. Primitive roots do not exist for all integers N† but they do exist for all primes. The proof may be found in almost any treatise on the theory of numbers [11]. A table of primitive roots for all primes less than 10^4 is given in the handbook of tables of Abramowitz and Stegun [1]. A given prime may have several primitive roots associated with it.

We illustrate the nature of the permutation with the prime $N = 11$, which has the primitive root $g = 2$.

i	1	2	3	4	5	6	7	8	9	10
$((g^i))$	2	4	8	5	10	9	7	3	6	1

Also, $((g^{N-1}))$ must always be 1, and therefore we can deduce that

$$((g^i)) = ((g^{i \bmod N-1})) \qquad (7.29)$$

† Primitive roots exist for $N = 2, 4$, and all powers of odd primes.

for all i. With this knowledge, we return to consideration of the DFT of a prime number of points, N [12].

The expression for F_0 is particularly simple,

$$F_0 = \sum_{n=0}^{N-1} f_n \tag{7.30}$$

and it will be considered separately later. We are anxious to compute i by a separate rule from the other F_k because it requires no multiplications We observe that for the other F_k the sample f_0 is not multiplied, and w choose to add it in last to the summations of (7.25). We are left then t consider the computation of the $(N-1)$-point sequence $\{F_k - f_0\}$ $k = 1, 2, \ldots, N - 1$, given by

$$F_k - f_0 = \sum_{n=1}^{N-1} f_n W^{nk} \tag{7.31}$$

Recalling the property of the primitive root g, we can allow the substitution

$$k \rightarrow ((g^k))$$

which merely rearranges the order of the $N - 1$ equations of (7.31), and the substitution

$$n \rightarrow ((g^n))$$

which simply reorders the terms in the summation. Therefore we can write

$$F_{((g^k))} - f_0 = \sum_{n=1}^{N-1} f_{((g^n))} W^{((g^n))((g^k))} \tag{7.32}$$

and therefore

$$F_{((g^k))} - f_0 = \sum_{n=1}^{N-1} f_{((g^n))} W^{((g^{(n+k)}))} \tag{7.33}$$

We are now able to recognize that the sequence $\{F_{((g^k))} - f_0\}$ is a cor-relation of the sequences $\{f_{((g^n))}\}$ and $\{W^{((g^n))}\}$ and, from (7.29), that it is an $(N-1)$-point circular correlation. Again, we can compute the DFT by the use of high-speed correlation.

There are two ways we might proceed. Since N is a prime, $N - 1$ is composite. If it is highly composite (for example, $N = 257$), then the $(N-1)$-point circular convolution may profitably be computed by the relation

$$\{F_{((g^k))} - f_0\} = \mathrm{DFT}^{-1}\{(\mathrm{DFT}\,\{f_{((g^{-i}))}\})(\mathrm{DFT}\,\{W^{((g^i))}\})\} \tag{7.34}$$

where all DFT operations called for are to be performed by an $(N - 1)$-point FFT algorithm. If $N - 1$ is highly composite, (7.34) is very effective.† However, if $N - 1$ is only modestly composite (as with $N = 563$), the savings from an FFT algorithm are overcome by the fact that more than one DFT has to be computed. However, a periodic correlation can always be computed as an aperiodic correlation based on DFTs using N' points, where N' is a highly composite number (say a power of 2) greater than $2N - 4$. Specifically we create a sequence $\{\gamma_n\}$ by inserting $N' - N + 1$ zeros between the zeroth and first points of the sequence $\{f_{((g^{-n}))}\}$, and we create a second sequence $\{\vartheta_n\}$ by periodically repeating $W^{((g^n))}$ until there are N' points. Then

$$\text{DFT}^{-1}\{(\text{DFT}\{\gamma_n\})(\text{DFT}\{\vartheta_n\})\}$$

contains $\{F_{((g^k))} - f_0\}$ as a subsequence: the first $N - 1$ points.

With either of these two procedures, about one-third of the computation may be saved if the transform of the sequence of powers of W has been precomputed. One method requires a computational effort proportional to $N - 1$ times the sum of the factors of $N - 1$, whereas the second method requires a computation proportional to $N' \log N'$. Furthermore, in the course of using either method we compute a DFT of the permuted input. The zeroth point of this DFT is the sum of the inputs, and if we add f_0 to this number we have eliminated the need to compute F_0 separately by (7.30). Still further, after we have multiplied the two DFTs together, part way through the method, we can accomplish the addition of f_0 to each of the other transform points by modifying the zeroth point of the product of DFTs; we add $(N - 1)f_0$ or $N'f_0$ to the zeroth point before inverse transforming, and the effect is to add f_0 to each output point, giving $\{F_{((g^k))}\}$, which needs only to be unscrambled to give the desired DFT of a prime number of points.

Note that the second method provided an example of the high-speed computation of a circular correlation where the number of points was not highly composite, as suggested by the footnote in Sec. 7.2.

7.9 HIGH-SPEED DIGITAL-FILTER DESIGN

A digital filter is said to be specified in the time domain if its impulse response β_n is determined. If high-speed techniques are to be applied, β_n must be nonzero in a single finite aperture, which may not always be the case. If it is not the case, high-speed techniques may be applied

† It is interesting that the $(N - 1)$-point DFT of $\{W^{((g^n))}\}$ has magnitude $\sqrt{N}$ at all points except the zeroth, where the DFT is equal to -1.

only approximately by forcing the β_n to assume zero value outside some judiciously chosen region. Generally speaking, the wider the region, the better is the approximation.

Region width is not the only important factor, however. Presumably the region must be positioned properly. The factors determining good positioning depend on the objectives of the filtering process, but common sense indicates that it is usually best to throw away values of β_n that are small and save those that are large.

There are a number of ways in which the β_n's might be determined. If the synthesis process is mathematical in nature, it is desirable to find ways of including the constraint of finite aperture on the process itself. In this way the filtering objectives can be coupled closely with the constraint, and a means of minimizing error measures might be devised.

If the β_n's are determined through physical measurements, as is often the case in problems involving simulation, the choices are less clear. Matters are best guided by considering the known physical properties of the measured object and attempting to preserve those that are important to the experiment.

For example, in an experiment involving the simulation of the acoustic properties of an auditorium, the impulse response between two chosen points can be determined by discharging electric sparks at one point and recording their effect at the other by means of a suitable microphone. Because of the physics of this situation, no contribution can be made to the β_n's until the effect of the sound of the spark has had a chance to reach the microphone. Since the distance is known, the value of n at which the beginning of the aperture should be placed can easily be computed. The determination of the aperture width is a little more difficult. Since the spark reverberations die out, it is possible to measure the physical properties of the room only over that interval for which the reverberations remain above measurement noise. Even if measurement noise is reduced indefinitely, the values of β_n decrease with increasing n and will eventually fall below the resolution of the A-D converter, thus defining an appropriate end for the aperture. If after these considerations, the aperture is still too large, its width might be reduced on the basis of the number of reverberation times that must be preserved or the fraction of total energy that can be thrown away without disturbing the validity of the simulation.

Whether the β_n's are determined mathematically or through measurements, once an appropriate aperture has been selected, the digital filter is completely specified and is ready for direct implementation via the methods of high-speed convolution.

A digital filter is said to be specified in the frequency domain if its frequency response $F(\omega T)$ is determined. $F(\omega T)$ is a continuous, periodic, omplex function of ωT, and for a nonrecursive filter it is given by

$$F(\omega T) = \sum_n \beta_n e^{-jn\omega T} \qquad (7.35)$$

ₛince $F(\omega T)$ is periodic it is completely specified once a single period is
ₚecified. It is customary to specify the period in the range $-\pi < \omega T$
$\leq \pi$. Since $F(\omega T)$ is complex, a real and imaginary part or a magnitude
ₙd phase must both be determined.

In the event that the β_n are real, as is usually the case, a constraint
ₛ placed upon $F(\omega T)$ which requires that the real part and magnitude of
ᵣ(ωT) be even functions of ωT and that the imaginary part and phase of
ᵣ(ωT) be odd functions. If this be the case, $F(\omega T)$ is completely specified
ₒnce it is specified in the region of $0 \leq \omega T \leq \pi$. A typical example is
ₛhown in Fig. 7.6 in polar form.†

Whatever the situation, once $F(\omega T)$ is specified, so are the β_n, which
ₐre given by

$$\beta_n = \frac{1}{2\pi} \int_{-\pi}^{\pi} F(\omega T)e^{+j\omega Tn}\, d(\omega T) \qquad (7.36)$$

ₗt is important to realize that the β_n are, in fact, the coefficients of the
ᵤomplex Fourier series for the periodic function $F(\omega T)$. Generally speak-
ₙng, there will be an infinite number of nonzero β_n's since for most periodic
ₜunctions an infinite number of Fourier harmonics are required for the
Fourier series expansion.

If high-speed techniques are to be applied, we again recall that β_n

† If $\beta_n = 0$ for $n < 0$ the digital filter is said to be realizable because its present output
ᵢs independent of future inputs. If this is so, the Re $F(\omega T)$ and Im $F(\omega T)$ are related
ᵦy the Hilbert transform [7], and once one is determined the other is specified. Under
ₜhese same circumstances, although the $|F(\omega T)|$ and the $\underline{/F(\omega T)}$ are not so tightly
related, there are certain conditions [7] that must be fulfilled by them.

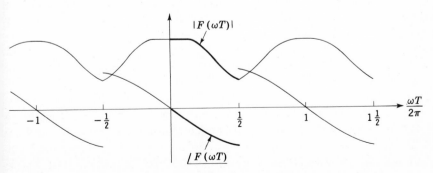

Fig. 7.6 Specification of frequency response for digital filter (in polar form).

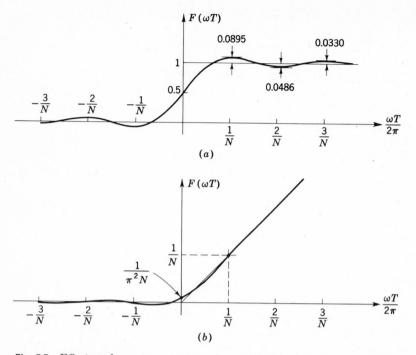

Fig. 7.7 Effect on frequency response of truncating impulse response: (a) in neighborhood of discontinuity in frequency response; (b) in neighborhood of discontinuity of derivative of frequency response.

must be nonzero in a single finite aperture. In the light of the preceding paragraph, we see that high-speed techniques can also be applied only approximately to the realization of digital filters specified in the frequency domain.

The nature of the approximations involved here are very well known since they are identical with those associated with the convergence of Fourier series. Central to this topic is the so-called Gibbs phenomenon, by now a very familiar concept, which manifests itself as a fixed-percentage overshoot and ripple before and after an approximated discontinuity. The characteristics of these ripples and those surrounding approximated discontinuous derivatives are obtained by studying the $(\sin x)/x$ function and its various integrals [7].

Figure 7.7 shows the effect of these approximations on $F(\omega T)$ in the vicinity of a unit discontinuity in value and a unit discontinuity in slope when β_n is limited to contain only N nonzero samples. It is important to notice that the ripples in Fig. 7.7a do not decrease with increasing N and that the first overshoot is about 9 percent of the size of the discontinuity. As N increases, the ripples squeeze into a narrower interval around the discontinuity, but their amplitudes remain the same, decreasing only in inverse proportion to their distance from the discontinuity. This means that the tenth overshoot is as large as 1 percent (-40 db) and at the hundredth the ripples still represent 0.1 percent (-60 db) of the size of the discontinuity.

In Fig. 7.7b the situation is slightly different. The maximum error occurs at the slope discontinuity in the form of a small "fillet." In contrast to the situation in Fig. 7.7a, the size of the fillet and the ripples surrounding it bear an inverse proportion to N. As N increases, the ripples squeeze into a narrower interval around the slope discontinuity and their amplitudes fade away at the same rate, the effect being the same as if Fig. 7.7b were viewed from an ever-increasing distance. In spite of their lesser significance, the ripples of Fig. 7.7b also decrease only in inverse proportion to their distance from the discontinuity.

Although a periodic function such as $F(\omega T)$ can require an infinite number of harmonics for its Fourier series representation even if it does not contain discontinuities in any derivative, the distortions of Fig. 7.7 are a fair representation of the difficulties to be encountered in high-speed digital filtering. This is brought about partly by the fact that a very large fraction of frequency-domain filters are required to have different arbitrary response characteristics in adjacent frequency bands, thus inducing a discontinuity at the boundaries between these bands. The best example of this is the ideal low-pass filter design for which

$$F(\omega T) = \begin{cases} 1 & 0 \le |\omega T| \le \omega_c T \\ 0 & \omega_c T < |\omega T| \le \pi \end{cases} \tag{7.37}$$

At $\omega_c T$ the approximate $F(\omega T)$ exhibits the same behavior as in Fig. 7.7a. When the frequency characteristics are shown on a decibel scale, as in Fig. 7.8, the penalties of the approximation are most clearly appreciated. Although the ripples that occur in the passband are almost unnoticeable (the largest represents about 0.75 db), those in the stopband are extremely troublesome, preventing the minimum attenuation from exceeding a little under 21 db. Most filtering applications require at least 40-db minimum attenuation, and it is not unreasonable to expect 60 to 80 db.

Even in situations that do not require frequency bands, as in the example just given, discontinuities in behavior are almost always required at $\omega T = 0$ and $\omega T = \pi$ (see Fig. 7.6). These points are, in effect, mirror points. At them the function $F(\omega T)$ is required to exhibit symmetry such that on opposite sides it assumes conjugate complex values. To avoid discontinuities the odd part of $F(\omega T)$ and any of its derivatives must possess a limit equal to zero at these points.† For $F(\omega T)$ and any

† Discontinuities in $\underline{/F(\omega T)}$ and the derivatives of $|F(\omega T)|$ do not necessarily indicate corresponding discontinuities in $F(\omega T)$. See, for example, Fig. 7.9.

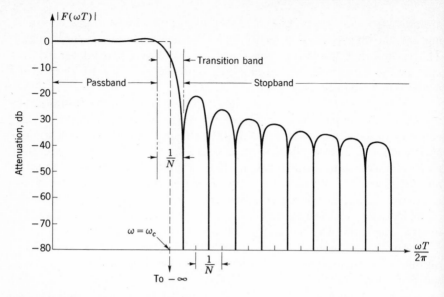

Fig. 7.8 Approximation to ideal low-pass filter, exhibiting Gibbs' phenomenon on decibel scale.

of its even-numbered derivatives the odd part corresponds to the imaginary part and for odd-numbered derivatives to the real part.

An excellent illustration of these effects is to be found in the perfect differentiator [10] for which

$$F(\omega T) = j\omega \qquad -\pi < \omega T < \pi \tag{7.38}$$

This $F(\omega T)$ is continuous at $\omega T = 0$ but possesses a jump of $2j\pi/T$ at $\omega T = \pi$. Figure 7.9 gives a plot of this $F(\omega T)$ and its approximation in both polar and cartesian form for $N = 32$.

A closely related $F(\omega T)$ exhibiting a discontinuity at $\omega T = 0$ as well as at $\omega T = \pi$ is that of the Hilbert transformer, which is used heavily for producing analytic functions and in single-sideband applications. Its frequency response is

$$F(\omega T) = \begin{cases} j & 0 < \omega T < \pi \\ -j & -\pi < \omega T < 0 \end{cases} \tag{7.39}$$

Figure 7.10 gives this $F(\omega T)$ and its approximation in a manner similar to that used for Fig. 7.9.

Ripples induced by slope discontinuities are most frequently encountered at $\omega T = 0$ or $\omega T = \pi$ and especially at the latter. The situation is ordinarily not serious unless $F(\omega T)$ also becomes small at the same points.

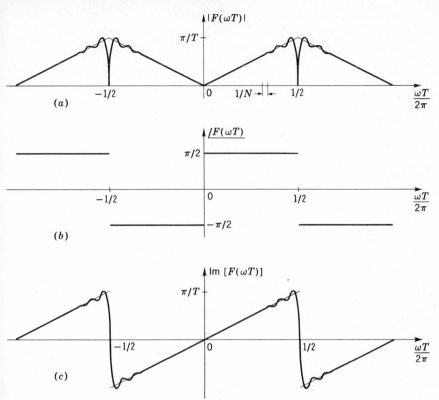

Fig. 7.9 Ideal differentiator and its approximation: (*a*) magnitude response; (*b*) phase response; (*c*) imaginary part of response.

An example of this is the Fejér or Cesàro filter which possesses a triangular frequency response given by

$$F(\omega T) = 1 - \left| \frac{\omega T}{\pi} \right| \qquad -\pi \le \omega T \le \pi \qquad (7.40)$$

This filter has a discontinuity in slope at $\omega T = 0$ where it is large and at $\omega T = \pi$ where it is small. Figure 7.11 illustrates this situation on a decibel scale. The error at $\omega T = 0$ and at $\omega T = \frac{1}{2}$ is approximately $4/\pi^2 N \approx \frac{1}{80}$. At $\omega T = 0$ this represents about 0.11 db, and at $\omega T = \pi$ it represents a minimum dip of about -38 db. Six decibels of attenuation may be added to the latter figure by doubling N.

Returning again to the consideration of frequency bands, we shall develop an example demonstrating the effect of slope discontinuities. Suppose we consider the Cesàro filter again but limit its bandwidth to a

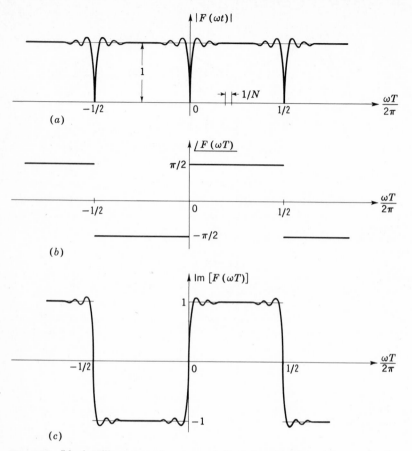

Fig. 7.10 Ideal Hilbert transformer and its approximation: (*a*) magnitude response; (*b*) phase response; (*c*) imaginary part of response.

cutoff $\omega_c T$ so that

$$F(\omega T) = \begin{cases} 1 - \left| \dfrac{\omega}{\omega_c} \right| & 0 \le |\omega T| \le \omega_c T \\ 0 & \omega_c T < |\omega T| \le \pi \end{cases} \tag{7.41}$$

Figure 7.12 applies to this situation. In contrast with the ideal low-pass filter of Fig. 7.8, this filter has stopband ripples that decrease by 6 db each time N is doubled. In this example, we chose

$$\omega_c = 5\Omega$$

where

$$\Omega = \frac{2\pi}{NT}$$

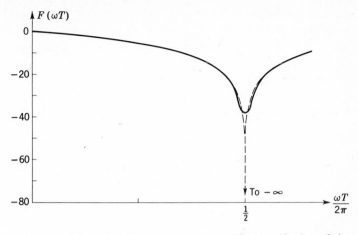

Fig. 7.11 Triangular frequency response (Cesàro filter) and its approximation on logarithmic scale.

The attenuation at $\omega T = \omega_c T$ is $\Omega/\pi^2\omega_c \approx \frac{1}{50}$, which represents -34 db. The largest stopband ripple is $0.34\Omega/\pi^2\omega_c \approx 0.0068$ or -43 db.

In view of these discussions and the computing time implied by (7.24) it would appear that the undesired ripples could be overcome simply by resorting to large enough N. This may be true for those encountered

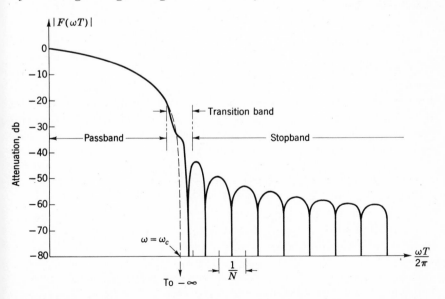

Fig. 7.12 Ideal Cesàro filter with cutoff frequency $\omega_c T$ (dashed line) and its approximation on logarithmic scale.

at a slope discontinuity such as in Fig. 7.12, because in that case for a fixed $\omega_c T$ value and transition bandwidth the minimum attenuation decreases 40 db for each decade increase in N. However, in the case of a step discontinuity as in Fig. 7.8, this solution is too costly even for high-speed digital methods.

The main stumbling block in these matters is the large overshoot of the Gibbs phenomenon and its slow decay. Fortunately a very large amount of study has been given to this problem and with considerable success [3, 10]. The theoretical considerations are extensive and sophisticated, with the most impressive results being obtained numerically by means of a digital computer [9, 10]. To summarize these efforts, suffice it to say that with the best techniques the stopband ripples of Fig. 7.8 can be controlled to provide a minimum attenuation of 80 db by multiplying N by 5 in conjunction with a weighting of the β_n within the aperture.

The idea of using a weighting function to modify the Fourier coefficients of a periodic function in order to control the convergence of that series is well known. It trades the rectangular window that limits an infinite series for a window of some other shape. Since the multiplication of Fourier coefficients by a window corresponds to convolving the periodic function with the Fourier transform of that window, efforts center around a search for a finite window the Fourier transform of which has relatively small side lobes.

The rectangular window has a transform of the form $(\sin x)/x$ which has relatively large side lobes, as we have seen. A triangular window has a transform of the form $[(\sin x)/x]^2$ which has much reduced side lobes that die out faster at the price of a doubly wide central lobe. A parabolic window can be found with a $[(\sin x)/x]^3$ transform, a triply wide central lobe, and only $\frac{1}{2}$ percent overshoot to a unit discontinuity. Since its side lobes decrease as the inverse cube of the distance from the central lobe, so do the ripples of a step discontinuity convolved with it. This situation represents a considerable improvement over that of Fig. 7.8. The minimum stopband attenuation is 46 db instead of 21 db, and the ripples fall off at the rate of 60 db per decade instead of 20 db per decade. Although these results are very good and easy to grasp, they do not reach the 80-db figure mentioned above.

Another edifying view into the suppression of ripples centers on the observation that the ripples alternate in sign. If they could be superimposed with a delay of one ripple, they would tend to cancel. In Fig. 7.13 two copies of Fig. 7.7a separated by $2\pi/N$ are shown separately and superimposed. The peak overshoot is reduced to 2 percent; since the ripples represent the first difference of those in Fig. 7.7a, they fall off as the inverse square of their distance from the discontinuity. The window

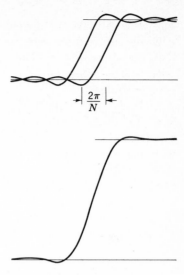

Fig. 7.13 Overcoming Gibbs' phenomenon by superimposing two shifted versions of frequency response to cancel ripples.

function corresponding to this modification is

$$w_n{}^{(1)} = \frac{1}{2}\left(e^{-j\pi n/N} + e^{+j\pi n/N}\right) = \cos\frac{\pi n}{N} \qquad -\frac{1}{2}N \le n \le +\frac{1}{2}N$$

$$(7.42)$$

and zero outside that range. This is shown in Fig. 7.14. Notice that β_n has been reduced for large values of n and that the transition band has been doubled in width.

Carrying the process one step further, we produce a superposition of basic step approximations which will form the second difference of those of Fig. 7.7a. To do this we require three copies separated by steps of $2\pi/N$, with the center one twice as large as the outer two. As a consequence, the peak overshoot is reduced to $\frac{1}{2}$ percent, and the ripples fall off as the inverse cube of their distance from the discontinuity. The window function corresponding is

$$w_n{}^{(2)} = \frac{1}{4}\left(e^{-j2\pi n/N} + 2 + e^{+j2\pi n/N}\right)$$

$$= \frac{1}{2}\left(1 + \cos\frac{2\pi n}{N}\right) \qquad -\frac{1}{2}N \le n \le \frac{1}{2}N \qquad (7.43)$$

Fig. 7.14 Window function in time domain; multiplying impulse response by this window function has effect of Fig. 7.13 in frequency domain.

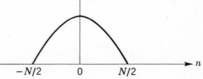

and zero outside the aperture. This window is known as the *hanning* window and is often mentioned in the literature [2, 3]. Similar to the $[(\sin x)/x]^3$ example cited above, it provides 44 db of stopband attenuation with ripples falling off at a rate of 60 db per decade. The second ripple is at -55 db.

Unlike the $[(\sin x)/x]^3$ example it is easy to see how to modify the $\frac{1}{4}$, $\frac{1}{2}$, $\frac{1}{4}$ weights of this example further to enhance the minimum stopband attenuation. The reason the first few overshoots are not completely canceled by their second differencing is that they are not a linear progression. The middle overshoot is always a little too small. To compensate for this, the central weight should be increased a small amount and the outside weights decreased. To null the first overshoot the correct weights are approximately 0.22, 0.56, 0.22; this agrees very closely with the *Hamming* window which is also discussed frequently [3]. A little further manipulation permits the first two ripples to be balanced off for a minimum of attenuation of about 60 db in the stopband.

These ideas may be carried further in an attempt to refine the cancellation of the ripples nearest the discontinuity. The additional weights become very small as in Blackman's window [3] (i.e., 0.04, 0.25, 0.42, 0.25, 0.04), producing a fivefold increase in transition bandwidth and a minimum attenuation of better than 87 db.

Although not for the purpose of canceling ripples one against the other, Kaiser's windows [9, 10] provide very fine refinements to the weights with which adjacent step approximations are superimposed so that very large stopband attenuations are compatible with minimum transition bandwidths. Whether this degree of refinement is required in a given application must be determined by a variety of factors, the most important of which is the upper limit on N as determined by the size of the available high-speed memory.

As a final example, Fig. 7.15 shows the effect obtainable with an appropriate window for the ideal low-pass filter first shown in Fig. 7.8. Notice the enlarged transition band and the increased attenuation in the stopband. The transition bandwidth in hertz is approximately $5/NT$. If we call λ the fraction of the total band (that is, $0 \leq \omega T \leq \pi$) used by the transition band, then

$$\frac{5}{N} = \frac{\lambda}{2} \tag{7.44}$$

and

$$N = \frac{10}{\lambda} \tag{7.45}$$

For $\lambda = 0.01$, $N = 1,000$ samples.

During the design of a high-speed digital filter it is profitable to use the FFT as a guide to the selection of an appropriate window function and

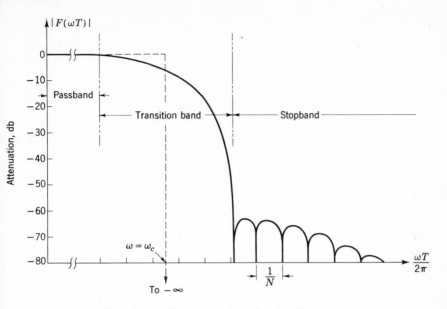

Fig. 7.15 Low-pass filter designed with aid of window function.

aperture size and position.   To accomplish this, $F(\omega T)$ should be computed in sampled form, using S samples, where S is a power of 2. S should be at least twice as large and preferably four or eight times as large as the initial estimate of N given by (7.45) in order to avoid alias error in computing β_n via the inverse FFT.   Once β_n is computed, an aperture of width N selected and positioned, and a window function chosen, the modified β_n should be formed by multiplication with the window.   Next, the DFT of the modified β_n should be computed to obtain samples of the modified $F(\omega T)$, still using S samples to provide resolution of detail.   A plot of these samples should be made for examination.   If the plot is unsatisfactory, the process should be repeated with a larger N, a different window, a different aperture position, or a combination of these.   When a satisfactory modified $F(\omega T)$ is obtained, the corresponding β_n can be used in a high-speed convolution to effect the filtering process. Of course, $\hat{\beta}_n$ must be formed on the basis of a period L with its nonzero values adjacent to $n = 0$.   Thus it is almost certain that the β_n will have to be circularly shifted and its DFT computed again.

An example of a design typifying these methods and involving a discontinuous $F(\omega T)$ will now be presented.   Consider an ideal low-pass-filter design for which $\omega_c T = 0.32\pi$ and a transition band fraction $\lambda = 0.15$ is desired.   Using (7.45) we obtain 67 for the initial estimate of N. Accordingly we select S as a power of 2 that is near $8N$, namely, $S = 512$. Next we compute $F(\omega T)$ in sampled form.   A plot is shown in Fig. 7.16a.

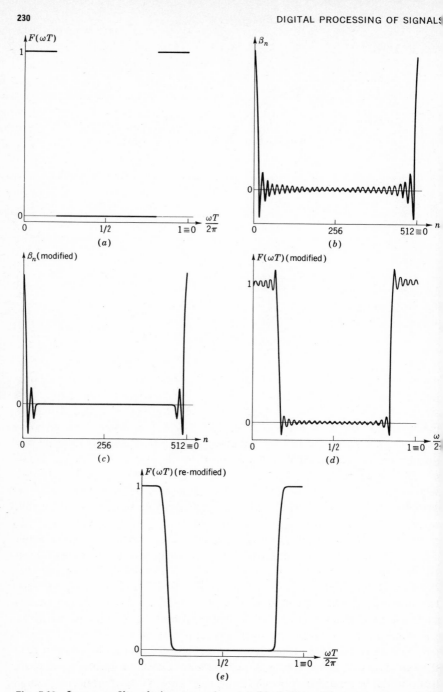

Fig. 7.16 Low-pass-filter-design example. (a) Ideal filter; (b) β_n as computed by 512-point FFT; (c) β_n of Fig. 7.16b multiplied by rectangular window; (d) frequency response corresponding to β_n of Fig. 7.16c; (e) frequency response corresponding to use of Blackman window.

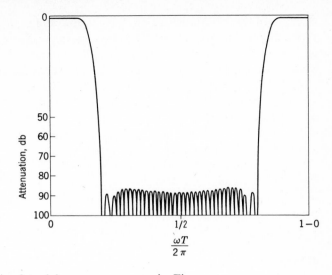

Fig. 7.17 Logarithmic plot of frequency response in Fig. 7.16e.

Now we calculate β_n via the inverse FFT, and the result is shown in Fig. 7.16b.   For purposes of illustration we first choose a rectangular window centered at $n = 0$ and of width $N = 65$.†   Multiplying β_n by this window we obtain the modified β_n shown in Fig. 7.16c.   When we compute the FFT of this modified β_n, we obtain samples of the modified $F(\omega T)$ shown in Fig. 7.16d.   This result is very ripply at the expense of a much faster transition bandwidth than required.   As a result we elect to use a Blackman window, which yields the modified $F(\omega T)$ of Fig. 7.16e.   This $F(\omega T)$ is also shown on a decibel scale in Fig. 7.17.   The minimum stopband attenuation is more than 87 db, and the transition band fraction λ is 0.172.   If N were increased slightly, λ could easily be brought to the design value of 0.15.   For an example of the use of this method involving slope discontinuities in $F(\omega T)$, see Helms [8].

7.10 SUMMARY

The product of DFTs is the DFT of a circular convolution.   This permits a circular convolution to be computed in a time proportional to $N \log N$ rather than N^2.   By appending appropriate numbers of zero-valued samples to each of the component functions of a circular convolution, it is possible to make the numerical values of the circular convolution identical to those of an aperiodic convolution.   An identical argument applies to correlation, which can be viewed as convolution with one sequence

† The rectangular window is 1 for $|n| < 32$, 0 for $|n| > 32$, and $\frac{1}{2}$ for $|n| = 32$, giving a total of 65 nonzero samples.

reversed in time.   Additional savings in time and in memory are available when only a subsequence of the results are required or when one sequence is very much longer than the other.   The former situation often arises in computing correlation functions for spectral estimation, and the latter situation is the case in nonrecursive digital filtering, for which one of the sequences is of indefinite length (arbitrarily long).   Among the applications of high-speed convolution or correlation are digital filtering, spectral estimation, and, surprisingly, the computation of discrete Fourier transforms that cannot be computed directly by the FFT.   The last application is possible because a DFT can be expressed as a convolution or correlation.   Considerable attention was given to the problems encountered in the design of nonrecursive digital filters and to the approach to be taken in overcoming these problems.   The FFT is a useful tool not only in realizing a nonrecursive filter but also in designing the filter.

REFERENCES

1. Abramowitz, M., and I. Stegun: "Handbook of Mathematical Functions," pp 827, 864–869, Dover Publications, Inc., New York, 1965.
2. Bingham, C., M. D. Godfrey, and J. W. Tukey: Modern Techniques of Power Spectrum Estimation, *IEEE Trans. Audio Electroacoustics*, **AU-15**(2): 56–66 (June, 1967).
3. Blackman, R. B., and J. W. Tukey: "The Measurement of Power Spectra," sec. B.5, pp. 95–100, Dover Publications, Inc., New York, 1958.
4. Bluestein, L. I.: "Several Fourier Transform Algorithms," *Nerem Record*, **10**: 218–219 (November, 1968).   Published by Boston section of Inst. of Electrical and Electronic Engineers.
5. Cooley, J. W., P. Lewis, and P. Welch: The Fast Fourier Transform Algorithm and Its Applications, *IBM Res. Paper RC-1743*, 1967.
6. Gentleman, W. M., and G. Sande: Fast Fourier Transforms—for Fun and Profit, 1966 Fall Joint Computer Conference, *AFIPS Proc.*, **29**: 563–578 (1966).
7. Guillemin, E. A.: "Theory of Linear Physical Systems," chap. 16, sec. 5, and chap 18, John Wiley & Sons, Inc., New York, 1963.
8. Helms, H. D.: Fast Fourier Transform Method of Computing Difference Equations and Simulating Filters, *IEEE Trans. Audio Electroacoustics*, **AU-15**(2): 85–90 (June, 1967).
9. Kaiser, J. F.: A Family of Window Functions Having Nearly Ideal Properties, unpublished memorandum, November, 1964.
10. Kaiser, J. F.: Digital Filters, in F. K. Kuo and J. F. Kaiser (eds.), "System Analysis by Digital Computer," John Wiley & Sons, Inc., New York, 1966.
11. McCoy, N.: "The Theory of Numbers," The Macmillan Company, New York, 1965
12. Rader, C.: Discrete Fourier Transforms When the Number of Data Samples Is Prime, *Proc. IEEE*, **56**: 1107–1108 (1968).
13. Singleton, R. C.: A Method for Computing the Fast Fourier Transform with Auxiliary Memory and Limited Highspeed Storage, *IEEE Trans. Audio Electroacoustics*, **AU-15**(2): 91–98 (June, 1967).
14. Stockham, T. G., Jr.: Highspeed Convolution and Correlation, 1966 Spring Joint Computer Conference, *AFIPS Proc.*, **28**: 229–233 (1966).

8
Generalized Linear Filtering
by Alan V. Oppenheim

8.1 INTRODUCTION

System theory and signal analysis have revolved primarily around spectral analysis and superposition. The fact that most waveforms can be expressed as a linear combination of complex exponential time functions and that complex exponential time functions are eigenfunctions of linear time-invariant systems has been the key to the representation of these systems. Thus linear-system theory has become a well-formulated and well-understood subclass of system theory in general. In characterizing linear systems it is possible to exploit their defining property, the principle of superposition. In contrast, when dealing with nonlinear systems one is faced with the problem that the general class is specified by the absence of a property. Consequently, it is impossible to develop a general characterization of nonlinear systems. The approach that must be taken, then, is to gather together those nonlinear systems that share common properties, with the hope that if these properties are carefully chosen they can be exploited in such a way as to lead to a useful characterization for that class of nonlinear systems.

There are many ways of classifying nonlinear systems, and each presents its difficulties. In this chapter one such classification is discussed. Specifically, nonlinear systems are represented in such a way that each class is defined by a principle of superposition, similar to the principle of superposition satisfied by linear systems [7]. Based on this representation, an approach to a class of nonlinear-filtering problems is discussed and developed in detail for the nonlinear filtering of multiplied signals and for a companion problem, the nonlinear filtering of convolved signals. Such problems arise in a variety of contexts. Four specific applications have been pursued, two of which are presented here, namely, dynamic-range compression and contrast enhancement of images and the analysis and processing of speech.

Much of the material in this chapter is the product of a collaboration of the author with Ronald W. Schafer and Thomas G. Stockham, Jr. In particular, the analysis presented in Secs. 8.5 to 8.9 represents the joint efforts of the author and Ronald Schafer. The material presented in Sec. 8.11, which treats the application to image processing, is borrowed entirely from Thomas Stockham, Jr.

The general theoretical framework, presented in Secs. 8.2 and 8.3, was originally formulated, and is presented here, without any particular bias toward either continuous or discrete waveforms. However, in the presentation of the analysis for the filtering of convolved signals, all the arguments are phrased in terms of discrete convolutions and discrete waveforms. Although formulation in these terms was originally motivated by the fact that the system was to be realized on a digital computer, it in fact turned out to be far simpler mathematically to study the problem in terms of sequences rather than continuous functions.

In addition to a strong reliance on the mathematics of discrete sequences and z-transforms as reflected in Secs. 8.5 to 8.9, the material in this chapter draws heavily on the results and attitudes of the previous chapters. The image-processing applications required two-dimensional linear filtering which was, of course, carried out digitally. Although not emphasized in the presentation, the technique used took full advantage of the methods of high-speed convolution, applied in two dimensions. The applications to problems in deconvolution required the explicit computation of high-resolution spectra. It is fortunate that, at about the time that applications were seriously being investigated, the Cooley-Tukey algorithm for computing Fourier transforms was being disclosed. This algorithm offered the possibility of carrying out deconvolution by the techniques presented in this chapter with practical and available computing power.

2 GENERALIZED SUPERPOSITION

The principle of superposition, as it is stated for linear systems, requires that, if T is the system transformation, then for any two inputs† $x_1(t)$ and $x_2(t)$ and any scalar c,

$$T[x_1(t) + x_2(t)] = T[x_1(t)] + T[x_2(t)] \tag{8.1}$$

and

$$T[cx_1(t)] = cT[x_1(t)] \tag{8.2}$$

From this definition it is clear that a system with transformation φ given by

$$\varphi[x(t)] = e^{x(t)} \tag{8.3}$$

is nonlinear. However,

$$\varphi[x_1(t) + x_2(t)] = \varphi[x_1(t)]\varphi[x_2(t)] \tag{8.4}$$

and

$$\varphi[cx_1(t)] = \{\varphi[x_1(t)]\}^c \tag{8.5}$$

and hence the transformation specified by (8.3) satisfies a form of superposition in the sense that its response to the linear combination of a set of inputs as specified by (8.4) and (8.5) is determined by the response to each of the inputs in the set. Similarly, if we consider a transformation ψ given by

$$\psi[x(t)] = [x(t)]^2 \tag{8.6}$$

then it is seen that

$$\psi[x_1(t)x_2(t)] = \psi[x_1(t)]\psi[x_2(t)] \tag{8.7}$$

and

$$\psi[x_1{}^c(t)] = \{\psi[x_1(t)]\}^c \tag{8.8}$$

The transformation of (8.6) can be said to satisfy a principle of superposition in the sense that its response to a product of inputs is the product of the individual responses. This suggests, then, a generalization of the principle of superposition, as it is stated for linear systems, that will encompass at least some nonlinear systems. To state this principle formally, let us consider a system with transformation φ and let $\{x(t)\}$ denote the collection of possible inputs and $\{y(t)\}$ denote the collection of possible outputs. Let

$$x_1(t) \; \bigcirc \; x_2(t)$$

† We shall assume that the signals involved are functions of time t. However, it is important for the reader to realize that there is nothing in the arguments to be presented that prevents the consideration of signals that are functions of space, frequency, or any other parameter. Neither is there any restriction to the consideration of one-dimensional signals.

Fig. 8.1 Representation of homomorphic system with input operation ○, output operation □, and system transformation ϕ.

denote the combination of any two inputs under an operation ○ (e.g., addition, multiplication, convolution, etc.) and let $y_1(t)$ □ $y_2(t)$ denote the combination of any two outputs under an operation □. Similarly, let $c \rfloor x(t)$ denote the combination of an input $x(t)$ with a scalar c and $c{:}y(t)$ denote the combination of an output $y(t)$ with a scalar c. Then the system can be said to satisfy a generalized principle of superposition if

$$\varphi[x_1(t) \bigcirc x_2(t)] = \varphi[x_1(t)] \;\square\; \varphi[x_2(t)] \tag{8.9}$$

and

$$\varphi[c \rfloor x(t)] = c{:}y(t) \tag{8.10}$$

For example, if φ is a linear transformation, the operations ○ and □ correspond to addition and the operations $\rfloor$ and : correspond to multiplication.

The basis for the characterization of systems having the properties of (8.9) and (8.10) lies in representing the system inputs as vectors in a vector space, the system outputs as vectors in a possibly different vector space, and the system transformation as a linear transformation between these spaces. With this approach, an additional restriction must be imposed on the operations ○ and $\rfloor$, namely, that they satisfy the same algebraic postulates as vector addition.† With this restriction and the assumption that the inputs constitute a vector space, with ○ and $\rfloor$ corresponding to vector addition and scalar multiplication, the theorems of linear algebra can be directly applied to the characterization of these systems. Such systems, which can be represented as linear transformations between vector spaces, have been referred to as homomorphic systems, a term suggested by the algebraic definition of a homomorphic (i.e., linear) mapping between vector spaces. The operation ○ is referred to as the input operation of the system, and the operation □ is referred to as the output operation. A homomorphic system with input operation ○, output operation □, and system transformation φ is represented as shown in Fig. 8.1.

The representation of homomorphic systems stems from the fact that under the stated restrictions on the set of inputs, an invertible system $\alpha_\bigcirc$ can always be found having the property that

$$\alpha_\bigcirc[x_1(t) \bigcirc x_2(t)] = \alpha_\bigcirc[x_1(t)] + \alpha_\bigcirc[x_2(t)]$$

and

$$\alpha_\bigcirc[c \rfloor x_1(t)] = c\alpha_\bigcirc[x_1(t)]$$

† See, for example, Ref. 4.

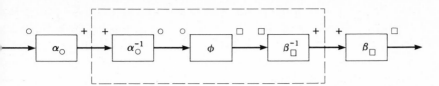

Fig. 8.2 Equivalent representation of homomorphic system.

.e., that it is homomorphic with input operation $\bigcirc$ and output operation $+$.

Similarly, we can find an invertible homomorphic system $\beta_\square$ with nput operation $\square$ and output operation $+$. Since $\alpha_\bigcirc$ and $\beta_\square$ are both nvertible, the system φ can then be represented as shown in Fig. 8.2. The system enclosed by dashed lines is a linear system; that is, it is a homomorphic system with addition as both the input and output operations. If L denotes this linear transformation, Fig. 8.2 can be redrawn as shown in Fig. 8.3. This cascade will be referred to as the canonic form or homomorphic systems. It is important to note that the system $\alpha_\bigcirc$ is determined only by the set of inputs and the input operations and that the system $\beta_\square$ (or $\beta_\square^{-1}$) is determined only by the set of outputs and the output operations. If we consider classifying homomorphic systems by their input and output spaces (including a specification of the input and output operations), then the systems within a class differ only in the linear portion of the canonic representation for that class. It can also be verified that the system of Fig. 8.3 is homomorphic with input operation $\bigcirc$ and output operation $\square$ for any choice for the linear system L. Consequently, when the characteristic systems $\alpha_\bigcirc$ and $\beta_\square$ for a class are known, the class of homomorphic systems can be generated by varying the linear system L. It can also be shown that if and only if the input and output operations for a class are memoryless (such as addition and multiplication in contrast with convolution) then the characteristic system for that class can be chosen to be memoryless. In these cases, then, all the system memory is concentrated in the linear portion of the canonic representation.

To help focus these ideas, let us consider the class of homomorphic systems with multiplication as the input operation and addition as the output operation. As a first step, we must verify that a vector space can

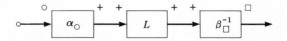

Fig. 8.3 Canonic representation of homomorphic system.

be constructed for which vector addition corresponds to the product of input functions. Let us restrict the class of possible inputs to positive time functions. Since multiplication is commutative and associative, it satisfies the same algebraic postulates as addition. For this choice of vector addition, the inverse of a vector corresponds to the reciprocal of the corresponding time function and, since our time functions are restricted to be positive, each input has a corresponding inverse. Restricting the set of scalars to be real, scalar multiplication in the vector space may be defined as a time function raised to a scalar power. In other words, if $x_1(t) \bigcirc x_2(t) = x_1(t)x_2(t)$, then $c \lfloor x(t) = [x(t)]^c$. It can be readily verified that this choice for scalar multiplication is consistent with the required algebraic postulates. The characteristic system $\alpha_\bigcirc$ can be derived by recognizing that, since the input operation is memoryless, the characteristic system is also memoryless. Thus, we wish to determine a system $\alpha_\bigcirc$ such that

$$\alpha_\bigcirc(x_1 x_2) = \alpha_\bigcirc(x_1) + \alpha_\bigcirc(x_2)$$

and

$$\alpha_\bigcirc(x_1{}^c) = c\alpha_\bigcirc(x_1) \tag{8.11}$$

From (8.11),

$$\alpha_\bigcirc{}^{-1}(cy_1) = [\alpha_\bigcirc{}^{-1}(y_1)]^c$$

where

$$y_1 = \alpha_\bigcirc(x_1)$$

Choosing y_1 equal to unity,

$$\alpha_\bigcirc{}^{-1}(c) = [\alpha_\bigcirc{}^{-1}(1)]^c$$

or, denoting $\alpha_\bigcirc{}^{-1}(1)$ by b,

$$\alpha_\bigcirc{}^{-1}(c) = b^c$$

so that

$$\alpha_\bigcirc(x_1) = \log_b x_1$$

Hence, the characteristic system for this case is a logarithmic amplifier, with the base of the logarithm arbitrary. The canonic form for this class of systems is then as shown in Fig. 8.4.

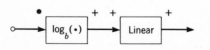

Fig. 8.4 Canonic form for class of homomorphic systems for which multiplication is input operation and addition is output operation.

8.3 GENERALIZED LINEAR FILTERING

In the design of filters, or, more generally, systems for waveform processing, linear systems have played an essential role. When we design the
linear portion of a waveform-processing system, we can often justify our
decisions and appeal to formal design procedures. In the nonlinear
portions, on the other hand, we are guided more frequently by intuition
and empirical judgments. The notion of generalized superposition
appears, at least in some cases, to offer a formalism for a class of nonlinear-
filtering problems which is an extension of the formalism that establishes
the basis for linear filtering.

The linear-filtering problem, as it is often stated, is concerned with
the use of a linear system for the recovery of a signal after it has been
added to noise. From a vector-space point of view, the linear-filtering
problem can be considered that of determining a linear transformation
on a vector space such that the length or norm of the error vector is
minimum. The norm associated with the vector space specifies the error
criterion to be used. There are many cases in which, when signal and
noise have been added, the best system to use is not a linear system. For
example, consider a signal that has been quantized with quantization
levels of 1, 2, 3, . . . , and assume that noise with peak values of ± 0.25
has been added to it. Clearly, the signal can be recovered exactly by use
of a quantizer, although there is no formal procedure that will suggest
this as the optimum nonlinear filter. In less obvious cases, there may
exist formal procedures both for a "best" linear filter and a "best" nonlinear filter from some class, although a meaningful and definitive
comparison between these filters cannot always be carried out, in part
because they often employ different information about the inputs.

It should be clear that a generalization can be carried out for the
filtering of signal and noise that have been nonadditively combined,
provided that the rule of combination satisfies the algebraic postulates of
vector addition. For example, if we wish to recover a signal $s(t)$ after it
has been combined with noise $n(t)$ such that the received signal is $s(t) \bigcirc n(t)$,
we may associate $s(t)$ and $n(t)$ with vectors in a vector space and the
operation $\bigcirc$ with vector addition. The class of linear transformations
on this vector space would then be associated with the class of homomorphic systems having the operation $\bigcirc$ as both the input operation and
the output operation. Hence, in generalizing the linear-filtering problem
to homomorphic filtering, the class of filters from which the optimum is
to be selected will be that class of homomorphic systems having input and
output operations that are identical to the rule under which the signals
that are to be separated have been combined. If x_1 and x_2 denote two
signals that have been combined under the operation $\bigcirc$, then the canonic
form for the class of homomorphic filters that would be used to recover

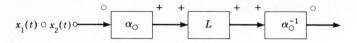

Fig. 8.5 Canonic form for class of homomorphic filters used to separate signals combined by operation $\bigcirc$.

x_1 or x_2 is that depicted in Fig. 8.5. The system $\alpha_\bigcirc$ and its inverse are characteristic of the class and hence, in selecting a system from the class, only the linear system L need be determined. Furthermore, we observe that, because the system $\alpha_\bigcirc$ is homomorphic with input operation $\bigcirc$ and output operation $+$, the input to the linear system L is $\alpha_\bigcirc(x_1) + \alpha_\bigcirc(x_2)$. Since the output of the linear filter is then transformed by means of the inverse of $\alpha_\bigcirc$, if the signal x_1 is to be recovered from the combination $x_1 \bigcirc x_2$, then the desired output from the linear system is $\alpha_\bigcirc(x_1)$. Consequently, the problem has been reduced to a linear-filtering problem, and all the formalism existing for that case can be applied here.

It must be emphasized that the approach to nonlinear filtering based on generalized superposition is but one possible approach among many. Its primary asset is that, just as with linear filtering of added signals, it is analytically convenient and, in fact, reduces to a linear-filtering problem. Although in most linear-filtering problems in practice we do not carry out a formal design procedure to determine the optimum choice, the error criterion that has received the most widespread attention is mean-square error (or integral-square error for aperiodic signals). In considering an error criterion for homomorphic filters, the natural choice would be the kind of error criterion that permits choosing the linear filter on the basis of mean-square error. This choice can be justified formally [8], but in any case it is natural to consider the system to be optimized when the linear filter is optimized.

Two kinds of problems in which the notion of homomorphic filtering has proved to be useful are in the filtering of multiplied signals and in the filtering of convolved signals. In the next sections these two classes of homomorphic filters will be developed.

8.4 HOMOMORPHIC FILTERING OF MULTIPLIED SIGNALS

There are a variety of problems in which a signal is expressible as a product of components that we wish to separate or modify separately. In transmission of a signal over a fading channel, for example, we may model the effect of fading in terms of a slowly varying component multiplying the transmitted signal, which we should like to remove at the receiver. As another example, an amplitude-modulated signal is represented by the product of a carrier signal and an envelope function, which we wish to separate at the receiver. Other examples include audio dynamic-range compression and image processing.

As has been discussed, multiplication satisfies the required algebraic postulates, and consequently we may consider using a homomorphic system. Thus we consider the class of filters described by the canonic form of Fig. 8.6 in which the system P has the property that

$$P[s_1 s_2] = P[s_1] + P[s_2] \qquad (8.12a)$$

and

$$P[s^c] = cP[s] \qquad (8.12b)$$

In general, we shall want to consider inputs that can be complex functions and associate a set of scalars with the inputs that are either real or complex. If we attempt to define as the characteristic system for this more general case a logarithmic transformation, we face the problem of resolving the ambiguity associated with the logarithm of a complex function. The standard artifice of invoking the principal value of the complex logarithm cannot be used in this case, because the principal value of the logarithm of a product of complex signals is not always the sum of the principal values corresponding to the individual complex signals, violating (8.12).

To specify the characteristic system for this more general case, we recall that the ambiguity in the complex logarithm function is a consequence of being permitted to add to the imaginary part of the complex logarithm arbitrary integer multiples of $j2\pi$; that is, we may add to the imaginary part a term of the form $jm(t)2\pi$, where $m(t)$ can assume only integer values. Roughly speaking, we can resolve this ambiguity if we require the imaginary part of the complex logarithm to be a continuous function of time passing through a specified point. Formally, let $\hat{s}(t)$ denote the response of the characteristic system P so that

$$\hat{s}(t) = P[s(t)]$$

We restrict $s(t)$ to be nonzero and continuous and define

$$\frac{d}{dt}\hat{s}(t) = \frac{1}{s(t)}\frac{ds(t)}{dt} \qquad (8.13)$$

or

$$\hat{s}(t) = \int_{t_0}^{t} \frac{1}{s(t)}\frac{ds(t)}{dt}\,dt \qquad (8.14a)$$

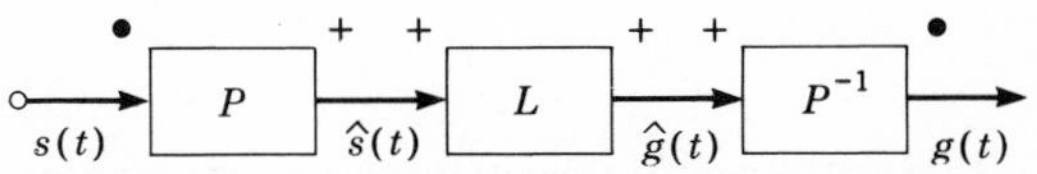

Fig. 8.6 Canonic form for class of homomorphic filters used to separate signals combined by multiplication.

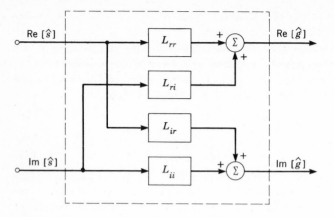

Fig. 8.7 General representation of linear system with complex input and output.

with

$$\hat{s}(t_0) = 0 \tag{8.14b}$$

Equations (8.13) and (8.14) specify an unambiguous choice for the definition of the complex logarithm; with this interpretation, we write that $\hat{s}(t) = \log s(t)$. The inverse system P^{-1} is the complex exponential function.

In general, we may permit the inputs and outputs of the system P to be complex and independently choose the set of scalars c associated with the inputs to be real or complex. In general, we may represent the linear system in the canonic form in terms of four subsystems, each of which has real inputs and outputs, as indicated in Fig. 8.7. If we restrict the scalars c to be real, then the system L of Fig. 8.6 need obey superposition only when its inputs are multiplied by real scalars. Under these conditions, the four subsystems of Fig. 8.7 are independent. If, instead, we associate complex scalars with the set of inputs, then we must require that the system L of Fig. 8.6 obey superposition when the inputs are multiplied by complex scalars. It can be verified in a straightforward manner that this requires that

$$L_{rr} = L_{ii} \tag{8.15a}$$

and

$$L_{ri} = -L_{ir} \tag{8.15b}$$

8.5 HOMOMORPHIC FILTERING OF CONVOLVED SIGNALS

Many waveforms of interest can be represented in terms of a convolution of component signals. In communicating or recording in a multipath or reverberant environment, for example, the effect of the distortion introduced can be modeled in terms of noise convolved with the desired signal. In speech processing, it is often of interest to isolate the effects of vocal-tract impulse response and excitation, which, at least on a short-time basis, can be considered to have been convolved to form the speech waveform [2]. Other examples lie in separation of probability-density functions that have been convolved by the addition of independent random processes.

A common approach to deconvolution is the technique of inverse filtering. In this case, the unwanted components of the signal to be processed are removed by filtering with a linear system whose system function is the reciprocal of the Fourier transform of these components. Clearly, this method is reasonable only for those situations in which there is a detailed model or description of the components to be removed. This approach has been successful, for example, in recovering the excitation function from the speech waveform since accurate models of the vocal tract have been developed. Inverse filtering is analogous to removing the effect of noise in the additive case (i.e., signal plus noise) by subtraction. If the noise is known exactly, except for a few parameters, one might reasonably expect to recover the signal by subtracting the noise from the sum. In many cases, however, detailed information about the unwanted components of the signal is not available, and consequently this method of subtraction in the additive case, or inverse filtering in the convolutional case, is no longer feasible.

In applying the notion of homomorphic filtering to the separation of convolved signals, we must first determine the characteristic system for this class of filters. Although the results may be formulated in terms of either continuous or discrete (sampled) inputs, the processing to be described is most easily realized on a digital computer. Consequently, the discussion will be phrased in terms of discrete time series.† Thus we consider a sequence $s(n)$ consisting of the discrete convolution of two sequences $s_1(n)$ and $s_2(n)$ so that

$$s(n) = \sum_{k=-\infty}^{+\infty} s_1(k)s_2(n - k)$$

or

$$s(n) = s_1(n) \otimes s_2(n) \tag{8.16}$$

where $\otimes$ denotes a discrete convolution. The canonic form for the class of filters is represented symbolically in Fig. 8.8, where D is the character-

---

† For notational convenience it will be assumed throughout the discussion that the sampling interval T is unity. Thus we represent samples of a time function as $f(n)$ rather than $f(nT)$. The z-transform variable z on the unit circle is represented by $z = e^{j\omega}$.

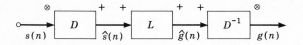

Fig. 8.8 Canonic form for class of homomorphic filters used to separate signals combined by convolution.

istic system for the class and has the property that

$$D[s_1(n) \otimes s_2(n)] = \hat{s}_1(n) + \hat{s}_2(n) \tag{8.17}$$

where $\hat{s}_1(n)$ and $\hat{s}_2(n)$ are the responses of D for inputs $s_1(n)$ and $s_2(n)$, respectively.

Let $S(z)$ and $\hat{S}(z)$ denote the two-sided z-transforms of $s(n)$ and $\hat{s}(n)$, respectively, so that

$$S(z) = \sum_{n=-\infty}^{+\infty} s(n)z^{-n} \tag{8.18a}$$

$$\hat{S}(z) = \sum_{n=-\infty}^{+\infty} \hat{s}(n)z^{-n} \tag{8.18b}$$

$$s(n) = \frac{1}{2\pi j} \oint_{c_1} S(z) z^{n-1} dz \tag{8.18c}$$

and

$$\hat{s}(n) = \frac{1}{2\pi j} \oint_{c_2} \hat{S}(z)z^{n-1} dz \tag{8.18d}$$

with c_1 and c_2 closed contours of integration in the z plane. It will be assumed for notational convenience that c_1 and c_2 are always taken to be the unit circle $z = e^{j\omega}$. Although this is somewhat restrictive, the results obtained are easily modified to incorporate the general case.

It follows from (8.16) and (2.12) that

$$S(z) = S_1(z)S_2(z) \tag{8.19}$$

where $S_1(z)$ and $S_2(z)$ are the z-transforms of $s_1(n)$ and $s_2(n)$, respectively. Hence, from the results of the previous section, applied here to functions of frequency, we may relate $S(z)$ and $\hat{S}(z)$ through a suitably defined logarithmic transformation.

Let us require that both $S(z)$ and $\hat{S}(z)$ be analytic functions with no singularities on the unit circle. Letting

$$S(e^{j\omega}) = S_R(e^{j\omega}) + jS_I(e^{j\omega}) \tag{8.20a}$$

and

$$\hat{S}(e^{j\omega}) = \hat{S}_R(e^{j\omega}) + j\hat{S}_I(e^{j\omega}) \tag{8.20b}$$

we then require that S_R, $\hat{S}_R$, S_I, and $\hat{S}_I$ be continuous functions of ω. Since the z-transform is a periodic function of ω with period 2π, we require in addition that $\hat{S}_R$ and $\hat{S}_I$ be periodic in ω. Furthermore, we may impose the constraint that $s(n)$ and $\hat{s}(n)$ be real functions so that S_R and $\hat{S}_R$ are even functions of ω and S_I and $\hat{S}_I$ are odd functions of ω. Then from (8.13) we define

$$\frac{d\hat{S}}{d\omega} = \frac{1}{S(\omega)} \frac{dS(\omega)}{d\omega}$$

so that

$$\hat{S}_R = \log |S| \qquad\qquad (8.21a)$$

and

$$\frac{d\hat{S}_I}{d\omega} = \frac{S_R{}^2}{S_I{}^2 + S_R{}^2} \frac{d}{d\omega} \frac{S_i}{S_r} \qquad\qquad (8.21b)$$

with

$$|\hat{S}_I(e^{j\omega})|_{\omega=0} = 0$$

Thus the imaginary part of $\hat{S}$ is interpreted to be the angle of S considered as a continuous, odd, periodic function of ω. The response of the system D then corresponds to the inverse transform of the complex logarithm of the transform.

A similar transformation was introduced by Bogert, Healy, and Tukey [1] in which the power spectrum of the logarithm of the power spectrum was proposed as a means for detecting echoes. The result of this set of operations was termed the cepstrum. It is clear that $\hat{s}(n)$ bears a strong relationship to the cepstrum with the primary differences being embodied in the use of the *complex* Fourier transform and *complex* logarithm. To emphasize the relationship while maintaining the distinction, it has been convenient to refer to $\hat{s}(n)$ as the complex cepstrum.

8.6 PROPERTIES OF THE COMPLEX CEPSTRUM

Although (8.21) defines the complex cepstrum, it is possible to reformulate the relationship between $s(n)$ and $\hat{s}(n)$ in several ways that place more in evidence the properties of the complex cepstrum. With (8.21) providing the interpretation of the complex logarithm, we write

$$\hat{s}(n) = \frac{1}{2\pi j} \oint \log [S(z)]z^{n-1} \, dz \qquad\qquad (8.22)$$

Since the contour of integration is the unit circle and we have defined $\log S(z)$ so that it is a single-valued function, we may rewrite (8.22) as

$$\hat{s}(n) = \frac{1}{2\pi} \int_{-\pi}^{\pi} \log [S(e^{j\omega})]e^{j\omega n} \, d\omega$$

Integrating by parts and using the fact that $S_I(e^{j\omega})$ is restricted to be a continuous, odd, periodic function of ω, we obtain the result

$$\hat{s}(n) = -\frac{1}{2\pi jn} \oint z \frac{S'(z)}{S(z)} z^{n-1} dz \qquad n \neq 0$$

$$\hat{s}(n) = \frac{1}{2\pi} \int_{-\pi}^{\pi} \log |S(e^{j\omega})| \, d\omega \qquad n = 0$$

(8.23)

where $S'(z)$ denotes the derivative of $S(z)$ with respect to z.

An example of a class of functions $S(z)$ satisfying the requirement that both $S(z)$ and $\hat{S}(z)$ are analytic is that class of the form

$$S(z) = |K| \frac{\displaystyle\prod_{i=1}^{M_0} (1 - a_i z^{-1}) \prod_{i=1}^{M_1} (1 - b_i z)}{\displaystyle\prod_{i=1}^{P_0} (1 - c_i z^{-1}) \prod_{i=1}^{P_1} (1 - d_i z)}$$

(8.24)

where the a_i and c_i are the zeros and poles inside the unit circle and $1/b_i$ and $1/d_i$ are the zeros and poles outside the unit circle. For this class of examples, we note that the poles of the integrand in (8.23) occur at the poles and zeros of $S(z)$. Consequently, $\hat{s}(n)$ will be composed of a sum of exponentials divided by n.

Equation (8.23) can be rewritten in a somewhat different form by noting that

$$\hat{S}'(z) = \frac{S'(z)}{S(z)}$$

or

$$\hat{S}'(z)S(z) = S'(z)$$

(8.25)

Using the fact that $z\hat{S}'(z)$ is the z-transform of $-n\hat{s}(n)$, and $zS'(z)$ is the z-transform of $-ns(n)$, the inverse z-transform of (8.25) is

$$[n\hat{s}(n)] \otimes s(n) = ns(n)$$

or

$$\sum_{k=-\infty}^{+\infty} \frac{k}{n} \hat{s}(k)s(n-k) = s(n) \qquad n \neq 0$$

(8.26)

In general, this is an implicit relation between $s(n)$ and $\hat{s}(n)$ and cannot be computed. However, if it is assumed that $s(n)$ and $\hat{s}(n)$ are zero for n negative and that $s(0) \neq 0$, then (8.26) becomes

$$\hat{s}(n) = \begin{cases} \dfrac{s(n)}{s(0)} - \displaystyle\sum_{k=0}^{n-1} \dfrac{k}{n} \hat{s}(k) \dfrac{s(n-k)}{s(0)} & n \neq 0 \\ \log s(0) & n = 0 \end{cases}$$

(8.27)

For this case, the inverse of the characteristic system can be easily obtained by solving (8.27) for $s(n)$ in terms of $\hat{s}(n)$ with the result

$$
s(n) = \begin{cases} s(0)\hat{s}(n) + \dfrac{1}{n} \displaystyle\sum_{k=0}^{n-1} k\hat{s}(k)s(n-k) & n \neq 0 \\[4mm] e^{\hat{s}(0)} & n = 0 \end{cases}
\tag{8.28}
$$

8.7 MINIMUM-PHASE SEQUENCES AND THE HILBERT TRANSFORM

For a certain class of functions, referred to as minimum-phase or minimum-delay functions, the phase associated with the Fourier transform is uniquely specified by the magnitude of the Fourier transform. For functions $f(t)$ of time with rational Laplace transforms, $f(t)$ is usually said to be minimum-phase if the Laplace transform contains no zeros or poles in the right half of the complex frequency plane. For sequences $f(n)$ with rational z-transforms, $f(n)$ is usually said to be minimum-phase if the z-transform contains no poles or zeros outside the unit circle.

For sequences (or time functions) that are minimum-phase, the log magnitude of the z-transform, evaluated on the unit circle, and the phase are related through the Hilbert transform. It is an interesting fact that the derivation of this relation can be interpreted in terms of the complex cepstrum. Specifically, as we shall see, a minimum-phase sequence can be defined as one whose complex cepstrum is zero for n (or t) less than zero. For minimum-phase sequences, then, the recursion relations of (8.27) and (8.28) apply. To develop these ideas in detail, it is helpful to review the notions of real-part sufficiency and Hilbert-transform relations as they apply to discrete sequences and z-transforms. The arguments parallel those presented in standard texts for the case of continuous time functions [3].

Consider a sequence $f(n)$ that is zero for $n < 0$ and whose z-transform evaluated on the unit circle is $F(e^{j\omega})$. We may express $f(n)$ as the sum of an even sequence and an odd sequence so that

$$f(n) = f_e(n) + f_o(n)$$

with

$$f_e(n) = \tfrac{1}{2}[f(n) + f(-n)]$$

and

$$f_o(n) = \tfrac{1}{2}[f(n) - f(-n)]$$

Since $f(n) = 0$ for $n < 0$, it follows that

$$f_o(n) = f_e(n)w(n) \tag{8.29}$$

and

$$f_e(n) = f_o(n)w(n) + f_e(0)\Delta(n) \tag{8.30}$$

where

$$w(n) = \begin{cases} 1 & n > 0 \\ 0 & n = 0 \\ -1 & n < 0 \end{cases}$$

If $F_R(e^{j\omega})$ and $F_I(e^{j\omega})$ denote the real and imaginary parts, respectively, of $F(e^{j\omega})$, it follows that $F_R(e^{j\omega})$ and $F_I(e^{j\omega})$ are the transforms of $f_e(n)$ and $f_o(n)$, respectively. Furthermore, from the complex convolution theorem of (2.16) and the relation (8.29),

$$jF_I(e^{j\omega}) = \frac{1}{2\pi} \int_0^{2\pi} F_R(e^{j\Omega}) G(e^{j(\omega - \Omega)}) \, d\Omega$$

where $G(e^{j\omega})$ denotes the transform of $w(n)$. It can be verified that

$$G(e^{j\omega}) = -j \cot \frac{\omega}{2}$$

and consequently

$$F_I(e^{j\omega}) = -\frac{1}{2\pi} \int_0^{2\pi} F_R(e^{j\Omega}) \cot \frac{\omega - \Omega}{2} \, d\Omega \qquad (8.31)$$

In a similar manner, it follows from (8.30) that

$$F_R(e^{j\omega}) = \frac{1}{2\pi} \int_0^{2\pi} F_I(e^{j\Omega}) \cot \frac{\omega - \Omega}{2} \, d\Omega + f_e(0) \qquad (8.32)$$

Equations (8.31) and (8.32) reflect the fact that, for sequences that are zero for $n < 0$, the imaginary part of the z-transform on the unit circle is uniquely specified by the real part and the real part is uniquely specified (to within a constant) by the imaginary part. This property is often referred to as real-part sufficiency, and (8.31) and (8.32) can be considered as the Hilbert-transform relationship between real and imaginary parts for discrete sequences.

Let us define a minimum-phase function as one for which the *phase* and *log magnitude* are related through the Hilbert transform so that

$$\theta(e^{j\omega}) = -\frac{1}{2\pi} \int_0^{2\pi} \log |F(e^{j\Omega})| \cot \frac{\omega - \Omega}{2} \, d\Omega$$

This then corresponds to applying the notion of real-part sufficiency to the logarithm of $F(e^{j\omega})$ or, equivalently, applying relations (8.29) and (8.30) to the complex cepstrum. Thus, with this interpretation, we may state equivalently that the complex cepstrum of a minimum-phase sequence is zero for $n < 0$.

To relate this interpretation to the statement that a minimum-phase sequence is one for which all poles and zeros lie inside the unit circle, we refer to (8.23). In general, the poles of the integrand occur at the poles and zeros of $S(z)$. If the unit circle is taken as the contour of integration,

$\hat{s}(n)$ will be zero for $n < 0$ if the integrand in (8.23) has no poles outside the unit circle, implying that $S(z)$ should have no poles or zeros outside the unit circle.

From the above discussion, we may conclude that for input sequences that are minimum-phase the input and output of the system D are related by the recursion relation of (8.27) and the input and output of the system D^{-1} are related by the recursion relation of (8.28). These recursion relations do not necessarily offer a computational advantage. However, they are conceptually important. In particular, they bring to light the fact that, for minimum-phase inputs, the transformation D is a realizable transformation; i.e., the response $\hat{s}(n)$ for $n = n_o$ is dependent only on samples of the input for $n \leq n_o$. Similarly, the inverse transformation D^{-1} is realizable for input sequences $\hat{s}(n)$ which are zero for $n < 0$. From this we may conclude that for minimum-phase input sequences the class of homomorphic filters defined by the canonic form of Fig. 8.8 is realizable in the sense that the output depends only on previous values of the input if the linear filter is also realizable.

An analogous discussion can be carried out for sequences whose complex cepstrum is zero for $n > 0$. Such sequences, which have no minimum-phase components, could appropriately be called maximum-phase sequences. For these cases a relation similar to (8.27) can be derived, in which values of the complex cepstrum depend only on future rather than past values of the input. It should be remarked that any sequence can always be expressed as the convolution of a minimum-phase sequence and a maximum-phase sequence, i.e.,

$$s(n) = s_1(n) \otimes s_2(n)$$

The portion of $\hat{s}(n)$ for $n > 0$ represents the contribution from the minimum-phase component, and the portion for $n < 0$ represents the contribution from the maximum-phase component.

From these considerations an interesting and perhaps useful result emerges. Consider a time-limited sequence $s(n)$ that contains $N + 1$ samples. Then its z-transform is a rational function with only zeros. Let us choose the origin and polarity of the waveform so that $S(z)$ can be expressed in the form of (8.24). Now $s(n)$ can be expressed as the convolution of a minimum-phase sequence $s_1(n)$ and a maximum-phase sequence $s_2(n)$, where $s_1(n)$ and $s_2(n)$ are time-limited so that

$$s_1(n) \begin{cases} \neq 0 & 0 \leq n \leq N_1 \\ = 0 & \text{elsewhere} \end{cases}$$

and

$$s_2(n) \begin{cases} \neq 0 & -N_2 \leq n \leq 0 \\ = 0 & \text{elsewhere} \end{cases}$$

where

$$N_1 + N_2 = N$$

The complex cepstrum of $s_1(n)$ is, in general, not time-limited. However, $\hat{s}_1(n)$ is zero for $n < 0$ and $\hat{s}_2(n)$ is zero for $n \geq 0$. Thus from (8.26)

$$s_1(n) = \begin{cases} e^{\hat{s}_1(0)} & n = 0 \\ \hat{s}_1(n)s_1(0) + \displaystyle\sum_{k=0}^{n-1} \left(\frac{k}{n}\right) \hat{s}_1(k)s_1(n-k) & n > 0 \end{cases}$$

and

$$s_2(n) = \begin{cases} 1 & n = 0 \\ \hat{s}_2(n) + \displaystyle\sum_{k=n+1}^{0} \left(\frac{k}{n}\right) \hat{s}_2(k)s_2(n-k) & n < 0 \end{cases}$$

Consequently, $N_1 + 1$ values of $\hat{s}_1(n)$ are needed to recover $s_1(n)$ and N_2 values of $\hat{s}_2(n)$ are needed to recover $s_2(n)$ so that $N_1 + N_2 + 1$ values of the complex cepstrum are needed to obtain the $N_1 + N_2 + 1$ values of $s(n)$.

8.8 SEQUENCES WITH RATIONAL z-TRANSFORMS

Thus far, we have restricted the input sequences to be such that $S(z)$ and $\hat{S}(z)$ are analytic and, for these cases, the logarithm of the z-transform on the unit circle was defined such that the imaginary part of the logarithm was a continuous, odd, periodic function of ω. It was remarked that this included all sequences with z-transforms of the form of (8.24). It is reasonable to assume that most input sequences of interest can be represented at least approximately by z-transforms which are rational, of the form

$$S(z) = K z^r \frac{\displaystyle\prod_{i=1}^{M_0} (1 - a_i z^{-1}) \prod_{i=1}^{M_0} (1 - b_i z)}{\displaystyle\prod_{i=1}^{P_1} (1 - c_i z^{-1}) \prod_{i=1}^{P_1} (1 - d_i z)} \tag{8.33}$$

Equation (8.33) differs from (8.24) in the inclusion of a term z^r representing a delay or advance of the sequence and in the removal of the absolute value on the multiplying constant so that $S(z)$ is no longer required to be positive for $z = 1$ ($\omega = 0$).

Although it is possible to generate a formal structure that would include this more general case, it offers no real advantage. Specifically, if we consider the problem at hand, namely, carrying out a separation of convolved signals, we would not expect to be able to determine, and most likely would not be interested in determining, how much of the constant K, including its sign, was contributed by each. Similarly, we could not expect to be able to determine how much of the net advance or delay r was contributed by each. In summary, we can expect to be generally interested in the shape of the components and not their amplitudes or time origin.

If we are willing to permit this flexibility, we can measure the algebraic sign of K and the value of r separately and then alter the input (or its transform) so that the z-transform is in the form of (8.24).

8.9 COMPUTATION OF THE COMPLEX CEPSTRUM

On the basis of the previous discussion, for general input sequences the computation of the complex cepstrum requires a computation of the Fourier transform of the input. Thus practical considerations require that the input $s(n)$ contain only a finite number of points, that is, be time-limited, and that the transform be computed only at discrete frequencies. Thus, in an implementation of the transformation D, we replace the z-transform and its inverse by the discrete Fourier transform pair (DFT) defined as

$$F(k) = \sum_{k=0}^{N-1} s(n) W^{nk}$$

and

$$s(n) = \frac{1}{N} \sum_{k=0}^{N-1} F(k) W^{-nk}$$

where

$$W = e^{-2\pi j/N}$$

Thus, the complex cepstrum computed by use of the DFT is given by

$$\hat{s}_d(n) = \frac{1}{N} \sum_{k=0}^{N-1} [\log S(k)] W^{-nk}$$

$$= \frac{1}{N} \sum_{k=0}^{N-1} [\log |S(k)| + j\theta(k)] W^{-nk} \tag{8.34}$$

with

$$S(k) = \sum_{k=0}^{N-1} s(n) W^{nk}$$

It is straightforward to verify that $\hat{s}_d(n)$ is an aliased version of $\hat{s}(n)$, that is,

$$\hat{s}_d(n) = \sum_{a=-\infty}^{+\infty} \hat{s}(aN + n)$$

The effect of the aliasing depends on the value chosen for the rate at which the spectrum is sampled or, equivalently, on the value of N. In many cases this is not a severe problem since, as we have seen in Chap. 6, relatively fast and efficient means for computing the discrete Fourier transform for large N exist.

The phase curve $\theta(k)$ can be computed by first computing the phase modulo 2π and then "unwrapping" it to satisfy the requirement that it be continuous and odd. Simple algorithms for doing this are easily generated provided that the frequency spacing of adjacent points is sufficiently small.

As an alternative to computing the complex cepstrum by means of (8.34), $\hat{s}(n)$ may be obtained by forming the ratio of the derivative of the spectrum and the spectrum, as suggested by (8.23). In particular, since samples of the derivative of the spectrum, denoted by $\tilde{S}(k)$, can be obtained by

$$\tilde{S}(k) = \sum_{n=0}^{N-1} nf(n) W^{nk}$$

we obtain a discrete approximation $\hat{s}_d(n)$ as

$$\hat{s}_d(n) = -\frac{1}{jn} \frac{1}{N} \sum_{k=0}^{N-1} \frac{\tilde{S}(k)}{S(k)} W^{-nk} \tag{8.35}$$

The complex cepstrum, computed on the basis of (8.35), differs somewhat from that computed from (8.34). The difference can be expressed by observing that $n\hat{s}_d(n)$ as represented by (8.35) is an aliased replica of $n\hat{s}(n)$, that is,

$$n\hat{s}_d(n) = \sum_{a=-\infty}^{+\infty} (aN + n)\hat{s}(aN + n)$$

or

$$\hat{s}_d(n) = \frac{1}{n} \sum_{a=-\infty}^{+\infty} (aN + n)\hat{s}(aN + n) \tag{8.36}$$

We note that, in general, the effect of the aliasing introduced by the use of (8.35) is more severe than that introduced by (8.34). On the other hand, use of (8.34) requires the explicit computation of the unwrapped-phase curve whereas use of (8.35) does not.

.10   APPLICATIONS OF HOMOMORPHIC FILTERING

n the preceding sections, we have discussed the analytical aspects of
homomorphic filtering in general and multiplicative and convolutional
iltering in particular.   In the remainder of this chapter some specific
applications resulting from this theory will be presented.   Four applica-
ions have been successfully pursued, two utilizing the ideas of homomorphic filtering of multiplied signals and two utilizing the ideas of
homomorphic filtering of convolved signals.   The first two concern
dynamic-range compression and expansion of audio waveforms [17] and
dynamic-range compression and contrast enhancement of images [13].
The second two deal with the removal of echoes in reverberated wave-
orms [15, 16] and the analysis and bandwidth compression of speech
11, 12, 14].   We shall restrict the discussion in this chapter to two of
these applications, namely, the dynamic-range compression and contrast
enhancement of images and the analysis and bandwidth compression
of speech.

.11   HOMOMORPHIC PROCESSING OF IMAGES

As argued by Stockham, image formation is predominantly a multiplicative process.   In a natural scene the observable brightness, as
recorded on the retina or on photographic film, can be considered as the
product of two components: an illumination function and a reflectance
'unction.   The illumination function describes the illumination of the
scene and can be considered to be independent of the scene.†   The
reflectance function encompasses the details of the scene and can be
considered to be independent of the illumination.   Thus we represent
an image as a two-dimensional spatial signal expressed in the form

$$I_{x,y} = i_{x,y} \cdot r_{x,y} > 0$$

where $I_{x,y}$ is the image, $i_{x,y}$ is its illumination component, and $r_{x,y}$ is its
reflectance component.   Negative brightness values are excluded on
physical grounds, and zero brightness values are excluded on practical
grounds.

  Two frequently occurring objectives in image processing are
dynamic-range compression and contrast enhancement.   The first of
these arises because we often encounter scenes with excessive light-to-
dark ratios resulting in a dynamic range too large to fit the available
medium, such as photographic film.   The solution is often to record a

_____

† The illumination is not independent of the geometry of the scene but is essentially
independent of the reflectance of the scene.

modified intensity $I'_{x,y}$ related to $I_{x,y}$ as

$$I'_{x,y} = I^{\gamma}_{x,y} \tag{8.37}$$

The parameter γ is well known to photographers who, by selecting from a variety of photographic materials and using shorter or longer development times for them, control its numerical value. When γ is chosen to be positive but less than unity, dynamic-range compression is achieved.

In the other problem, the objective is to process the image in such a way that contrast is enhanced, lending more sharpness to the edges of objects. This contrast enhancement is often achieved by modifying the spatial brightness pattern according to (8.37), with γ chosen to be greater than unity.

With this specific approach it is clear that dynamic-range compression and contrast enhancement are conflicting objectives. Dynamic-range compression achieved by using a γ less than unity tends to reduce contrast and, if carried to extremes, gives the image a muddy or washed out appearance. Contrast enhancement achieved by using a γ of greater than unity increases the dynamic range of the image, often with the result that dynamic-range capacities are exceeded even more than before.

To a reasonable approximation, dynamic-range compression can be treated as a problem focusing on the illumination function $i_{x,y}$, and contrast enhancement can be treated as a problem focusing on the reflectance function $r_{x,y}$. That is, the large dynamic ranges encountered in natural images are contributed to mostly by large variations in illumination whereas the edges of objects contribute only to the reflectance component. Thus we can consider separating the reflectance and illumination functions, modifying each with different gammas, and then recombining to form the modified image. With this approach, the modified intensity $I''_{x,y}$ is given as

$$I''_{x,y} = (i_{x,y})^{\gamma_i}(r_{x,y})^{\gamma_r} \tag{8.38}$$

where γ_i is less than unity to compress dynamic range and γ_r is greater than unity to enhance contrast.

With this objective in mind, we may consider achieving the image processing with a homomorphic filter directed toward carrying out separate processing on the illumination and reflectance components. The form for this image processor would be as indicated in Fig. 8.9. The illumination function generally varies slowly whereas reflectance often (but not always) varies rapidly because objects vary in texture and size and almost always have well-defined edges. If $\log i_{x,y}$ and $\log r_{x,y}$ were to possess frequency components occupying separate spatial-frequency areas, it would be straightforward in terms of the system of Fig. 8.9 to process these components separately. It is reasonable to assume that

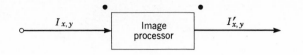

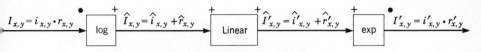

ig. 8.9 Canonic form for image processor directed toward modifying separately
illumination and reflectance components of original image.

og $i_{x,y}$ contributes mostly low spatial frequencies. The rapid variations
in $r_{x,y}$ contribute to the high spatial frequencies in log $r_{x,y}$ although the
reflectance also provides some contribution to the low spatial frequencies.
Thus, only partially independent processing is possible. Nevertheless,
t has proved to be reasonable in practice to associate the low spatial
frequencies in the logarithm with log $i_{x,y}$ and the high spatial frequencies
with log $r_{x,y}$. With this assumption, and with the desire to process the
image according to (8.38), the linear filter of Fig. 8.9 is chosen to multiply
he low spatial frequencies by γ_i and the high spatial frequencies by γ_r.

In implementing this processing, the filter frequency characteristic
was chosen to have the general shape of Fig. 8.10 and to be isotropic with
zero phase. The linear processing was performed through the use of
high-speed convolution methods, as discussed in Chap. 7, applied in two
dimensions. Two examples of images processed in this way for simultaneous dynamic-range compression and contrast enhancement are indicated in Fig. 8.11. The low-frequency gain of the filter was chosen to be
0.5, and the high-frequency gain was chosen to be 2, corresponding to a
choice of $\gamma_i = 0.5$ and $\gamma_r = 2$ in (8.38). Figure 8.11a shows the original
scenes, and Fig. 8.11b shows the processed scenes.

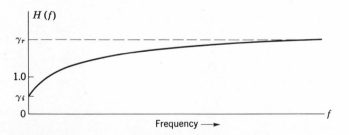

Fig. 8.10 Frequency characteristic used for linear filter of Fig. 8.9
when simultaneous dynamic-range compression and contrast
enhancement are to be achieved.

Fig. 8.11a Two original images.

Fig. 8.11b Images of Fig. 8.11a after processing to achieve simultaneous dynamic-range compression and contrast enhancement.

8.12 HOMOMORPHIC PROCESSING OF SPEECH

During voiced sounds such as vowels, the speech waveform may be considered the result of periodic puffs of air released by the vocal cords exciting an acoustic cavity, the vocal tract [2].   Thus a simple and often useful model of the speech waveform consists of the convolution of three components, representing pitch, the shape of the vocal cord or glottal excitation, and the configuration of the vocal tract.   Many systems for compressing the bandwidth of speech and for carrying out automatic speech recognition have as the basic strategy the separate isolation and characterization of the vocal-tract excitation and the vocal-tract impulse response.   Thus many speech-processing systems are directed in part toward carrying out a deconvolution of the speech waveform.

The speech waveform is a continuing signal and therefore must be processed on a short-time basis.   Thus we consider a portion $s(t)$ of the speech waveform as viewed through a time-limited window $w(t)$. Although the vocal-tract configuration changes with time, we shall choose the duration of the window to be sufficiently short so that we can assume that over this duration the shapes of the vocal-tract impulse response and the glottal pulse are constant.   Then, if we denote by $p(t)$ a train of ideal impulses whose timing corresponds to the occurrence of the pulses released by the vocal cords, by $g(t)$ the shape of the glottal pulse, and by $v(t)$ the impulse response of the vocal cavity, we express $s(t)$ approximately as

$$s(t) = [p(t) \otimes g(t) \otimes v(t)]w(t) \tag{8.39}$$

Furthermore, if $w(t)$ is smooth over the effective duration of the glottal pulse and the vocal-tract impulse response, we can approximate Eq. (8.39) as

$$s(t) = [p(t)w(t)] \otimes g(t) \otimes v(t) \tag{8.40}$$

Thus, if a smooth window is used to weight the speech waveform, we can consider the weighted aperiodic function as a convolution of weighted pitch, glottal pulse, and vocal-tract impulse response.

In keeping with the previous discussion, we wish to phrase our remarks in terms of samples of $s(t)$, which we denote by $s(n)$.   Assuming that we can replace the continuous convolution of (8.40) by a discrete convolution of samples of each of the component terms, we write

$$s(n) = [p(n)w(n)] \otimes g(n) \otimes v(n)$$

or

$$s(n) = p_1(n) \otimes g(n) \otimes v(n) \tag{8.41}$$

where $w(n)$, $g(n)$, and $v(n)$ are samples of $w(t)$, $g(t)$, and $v(t)$, respectively, and $p_1(n)$ is a train of unit samples weighted with the window $w(n)$.

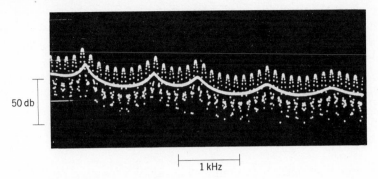

Fig. 8.12 Log spectrum of synthesized vowel.

Roughly speaking, the key to the separation of the pitch component $p_1(n)$ from the combined effects of vocal tract and glottal pulse lies in he fact that $p_1(n)$ tends to contribute rapid periodic variations to the og spectrum whereas $g(n) \otimes v(n)$ tends to contribute slow variations to he log spectrum.   This is illustrated in Fig. 8.12 which represents the og spectrum for a synthesized vowel.   The solid line represents the log pectrum of $v(n)$.   The rapid periodic variations superimposed on this epresent the log spectrum of $p_1(n)$.   On this basis we might expect that he inverse transform of the log spectrum would contain the contributions rom each of these components in different time regions.

More formally, the vocal-tract impulse response $v(n)$ can be modeled as the response of a cascade of damped resonators so that its z-transform is

$$V(z) = \frac{k}{\displaystyle\prod_{i=1}^{M}(1 - a_i z^{-1})(1 - a_i^* z^{-1})} \qquad |a_i| < 1 \qquad\qquad (8.42)$$

In this case, $v(n)$ is minimum-phase, and it follows from (8.23) that $\hat{v}(n)$, the complex cepstrum of $v(n)$, is of the form

$$\hat{v}(n) = \begin{cases} \displaystyle\sum_{i=1}^{M} \frac{|a_i|^n}{n} \cos \omega_i n & n > 0 \\[2mm] 0 & n < 0 \end{cases} \qquad\qquad (8.43)$$

where $a_i = |a_i| e^{j\omega_i}$.   Thus $\hat{v}(n)$ decays as $1/n$ and therefore tends to have its major contribution near the origin for $n > 0$.

An accurate analytical representation of the glottal pulse $g(n)$ is not known and consequently it is difficult to make any specific statements regarding the characteristics of its complex cepstrum $\hat{g}(n)$.   However, we can expect in general that $g(n)$ is non-minimum-phase [5].   Expressing

$g(n)$ as the convolution of a minimum-phase sequence $g_1(n)$ and a maximum-phase sequence $g_2(n)$, we shall assume that $\hat{g}_1(n)$, which is zero for $n < 0$, and $\hat{g}_2(n)$, which is zero for $n > 0$, both have an effective duration less than a pitch period.

The complex cepstrum of a train of weighted unit samples representing pitch is a train of unit samples with the same spacing. Specifically, if we let $P_1(z)$ denote the z-transform of $p_1(n)$, then

$$P_1(z) = \sum_{k=-\infty}^{+\infty} w(kN)(z^N)^{-k}$$

where N represents the number of samples between pitch pulses.

Letting $w_N(n)$ be defined as the window compressed in time by N so that $w_N(n) = w(nN)$, then

$$P_1(z) = w_N(z^N)$$

and

$$\log P_1(z) = \log w_N(z^N)$$

where $w_N(z)$ represents the z-transform of $w_N(n)$. Then $\hat{p}_1(n)$, the complex cepstrum of $p_1(n)$, is given by

$$\hat{p}_1(n) = \begin{cases} \hat{w}_N\left(\dfrac{n}{N}\right) & n = 0,\ \pm N,\ \pm 2N,\ \ldots \\ 0 & \text{otherwise} \end{cases} \tag{8.44}$$

In other words, the complex cepstrum of a train of pitch samples weighted with a window can be determined by compressing the window by a factor N corresponding to the spacing between pitch samples, determining the complex cepstrum, and expanding the result by a factor N. We observe that $\hat{p}_1(n)$ as expressed by (8.44) consists of a train of samples with spacing τ in which the mth sample has a weighting $\hat{w}_\tau(m)$.

The components of $s(n)$ due to pitch and due to the combined effects of vocal tract and glottal pulse tend to provide their primary contributions in nonoverlapping time intervals. The degree of separation will, of course, depend to some extent on the pitch, with more separation for low-pitched male voices than for high-pitched female voices. Experience has indicated, however, that except in cases of very high pitch a good separation of these components occurs. To illustrate, consider the example of Fig. 8.13. Figure 8.13a shows a portion of the vowel "ah," as in "father," with a male speaker, and Fig. 8.13b shows the complex cepstrum. In accordance with the previous discussion, we can recover the term $p_1(n)$ of (8.41) by multiplying the complex cepstrum by zero in the vicinity of the origin (with a time width of, say, 8 msec) and by unity elsewhere. Alternatively, to recover $v(n) \otimes g(n)$ we could multiply the complex cepstrum by unity in the vicinity of the origin and by zero elsewhere. After

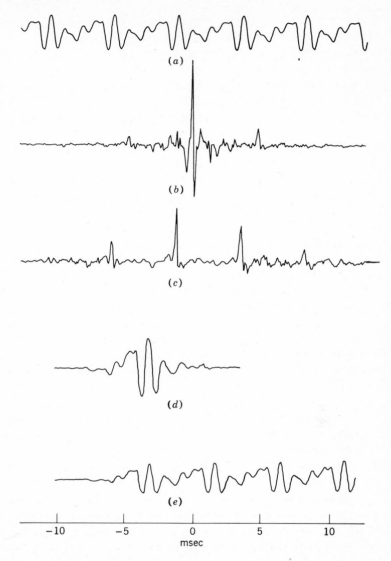

Fig. 8.13 (*a*) Portion of vowel. (*b*) Complex cepstrum of (*a*). (*c*) Recovered train of weighted pitch pulses. (*d*) Recovered vocal-tract impulse response. (*e*) Resynthesized speech using impulse-response function of (*d*) and pitch as measured from (*c*).

this weighting, the result is transformed by means of the system D^{-1}. Figure 8.13*c* shows the result of attempting to recover the weighted train of pitch pulses $p_1(n)$. Pulses with the correct spacing are clearly evident.†

---

† Pitch detection based directly on a measurement of the location of a peak in the cepstrum (as defined in [1]) has been successfully demonstrated by Noll [6].

In Fig. 8.13d the result of retaining only the low-time portion of $\hat{s}(n)$ corresponding to attempting to recover $v(n) \otimes g(n)$ is shown.   To verify that the pulse of Fig. 8.13d can be considered a convolutional speech component, the speech was resynthesized by convolving this pulse with a train of unit samples whose spacing was chosen to be a pitch period as measured from the waveform of Fig. 8.13c.   The resynthesized speech is shown in Fig. 8.13e and should be compared with the original speech of Fig. 8.13a.

These ideas have been applied to the simulation of a speech analysis-synthesis system directed toward studying the properties of speech and toward speech bandwidth compression [14].   The essential strategy is to analyze the speech into pitch and voiced-unvoiced information and information representing the combined glottal pulse and vocal-tract impulse response.   In the analysis, the cepstrum is obtained by weighting the input speech with a *hanning* window 40 msec in duration.   The cepstrum is separated into its high- and low-time parts with the cutoff time 3.2 msec.   A decision as to whether the 40-msec sample is voiced or unvoiced and a measurement of the pitch frequency if voiced are made from the high-time portion.   The synthesizer receives the low-time cepstral values, a voiced-unvoiced decision, and a pitch-frequency measurement updated at 10- or 20-msec intervals.   In the synthesizer, the pitch and voiced-unvoiced information are converted to an excitation function consisting of impulses during voicing and noise during unvoicing.   The low-time cepstral values are transformed back to an impulse-response function which is then convolved with the excitation function to form the synthetic speech.

In directing the system toward bandwidth reduction, the assumption was made that the impulse-response function is minimum-phase, thereby reducing by a factor of 2 the amount of information transmitted about the impulse-response function, since only the even part of the low-time cepstral values was retained.

Although it is generally accepted that speech is not minimum-phase, experiments with this system indicate that high-quality natural-sounding speech can be obtained by assuming minimum-phase.   The phase curve was generated in the synthesizer in accordance with the discussion of Sec. 8.7.   Figure 8.14 shows spectrograms of a sentence before and after processing.

8.13 SUMMARY

In this chapter the concept of generalized linear filtering was introduced and applied to the processing of signals that can be expressed as a product or as a convolution of component signals.   The basis for the approach stems from a realization that the principle of superposition can be gen-

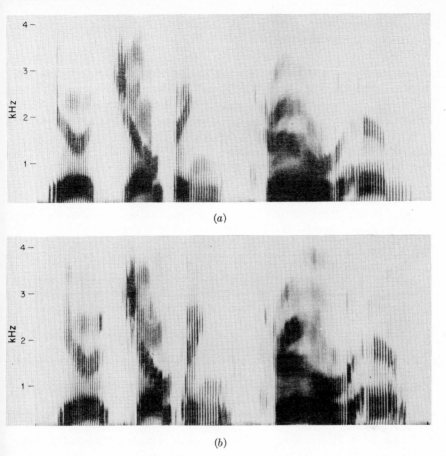

Fig. 8.14 (*a*) Spectrogram of sentence "your jumping thrilled him" as spoken by a male; (*b*) sentence of (*a*) after processing.

eralized in such a way that it applies to a broad class of nonlinear systems.

After the notion of generalized linear filtering was introduced, the specific considerations for the filtering of multiplied signals and the filtering of convolved signals were presented, and the relationship between the operations required for deconvolution and cepstral analysis were described. The properties of the complex cepstrum for minimum-phase sequences were developed, and the procedure for explicitly evaluating the Hilbert-transform relationship between spectral magnitude and phase by means of the complex cepstrum was described. Since the realization of these filters for deconvolution requires, in general, the determination of the spectrum, the properties and evaluation of the DFT play a central role in the discussion.   Section 8.9 in particular describes some of the practical details involved.

Two applications were described.   The first involves the dynamic-range compression and contrast enhancement of images, considered as a problem in the filtering of a signal described as a product of components. The second involves the analysis and bandwidth compression of speech viewed as a problem in the filtering of a signal described as a convolution of components.   Other applications which have been developed or appear to warrant consideration are mentioned.

REFERENCES

1. Bogert B., M. Healy, and J. Tukey: The Quefrency Alanysis of Time Series for Echoes, in M. Rosenblatt (ed.), "Proceedings of the Symposium on Time Series Analysis," chap. 15, pp. 209–243, John Wiley & Sons, Inc., New York, 1963.
2. Flanagan, J. L.: "Speech Analysis, Synthesis and Perception," Academic Press Inc., New York, 1965.
3. Guillemin, E. A.: "Theory of Linear Physical Systems," chap. 18, John Wiley & Sons, Inc., New York, 1963.
4. Hoffman, K., and R. Kunze: "Linear Algebra," Prentice-Hall, Inc., Englewood Cliffs, N.J., 1961.
5. Mathews, M. V., J. E. Miller, and E. E. David, Jr.: Pitch Synchronous Analysis of Voiced Sounds, J. Acoust. Soc. Am., 33(2): 179–186 (February, 1961).
6. Noll, A. M.: Cepstrum Pitch Determination, J. Acoust. Soc. Am., 41: 293–309 (February, 1967).
7. Oppenheim, A. V.: Superposition in a Class on Non-linear Systems, Mass. Inst. Technol., Res. Lab. Electron., Tech. Rept. 432, Mar. 31, 1965.
8. Oppenheim, A. V.: Optimum Homomorphic Filters, Mass. Inst. Technol., Res. Lab. Electron., Quart. Progr. Rept. 77, pp. 248–260, Apr. 15, 1965.
9. Oppenheim, A. V.: Non-linear Filtering of Convolved Signals, Mass. Inst. Technol., Res. Lab. Electron., Quart. Progr. Rept. 80, pp. 168–175, Jan. 15, 1966.
10. Oppenheim, A. V.: Generalized Superposition, Inform. Control, 11: 528–536 (November, 1967).
11. Oppenheim, A. V.: Deconvolution of Speech, J. Acoust. Soc. Am., 41(6): 1595 (abstract) (1967).
12. Oppenheim, A. V., and R. W. Schafer: Homomorphic Analysis of Speech, IEEE Trans. Audio Electroacoustics, AU-16: 221–226 (June, 1968).
13. Oppenheim, A. V., R. W. Schafer, and T. G. Stockham, Jr.: The Non-linear Filtering of Multiplied and Convolved Signals, Proc. IEEE, 56(8): 1264–1291 (August, 1968).
14. Oppenheim, A. V.: A Speech Analysis-synthesis System Based on Homomorphic Filtering, J. Acoust. Soc. Am., February, 1969 [to appear].
15. Schafer, R. W.: Echo Removal by Generalized Linear Filtering, Nerem Record, pp. 118–119, 1967.
16. Schafer, R. W.: Echo Removal by Discrete Generalized Linear Filtering, Ph.D. thesis, Massachusetts Institute of Technology, Department of Electrical Engineering, Cambridge, Mass., February, 1968.
17. Stockham, T. G., Jr.: The Application of Generalized Linearity to Automatic Gain Control, IEEE Trans. Audio Electroacoustics, AU-16: 267–270 (June, 1968).

Index

Accuracy (*see* Quantization effects)
Advantage of general purpose computer
    display, 9–13
Aliasing, 20–21, 138–140, 252
All-pass network, 90–92
Amplitude-modulated signal, 240
Analog-to-digital converter, 21
    noise caused by, 103–105
Aperiodic convolution, 205–206
Approximation problem, 50
Assemblers with macrodefinitions, 153
Audio dynamic-range compression,
    241, 253
Autocovariance, 121

Bandpass filters, 76–77, 83–86
Bandpass limiter, 138–140
Bandpass transformation, 76–78
Bank of filters, 84
Bessel filters, 56
Bilinear transform (*see* Digital filtering)
Biomedical research, 1, 8
Bit reversal, 177–182, 185–186, 197
Block diagram, 132–136
Brain waves, 9
Butterworth filters, 51, 56–59, 66–73

Calling sequence, 155–158
Canonic form, 36–37, 42–43
    for deconvolution, 243
    for homomorphic systems, 237, 241
    inadvisability of implementation, 103
Cardiograms, 9
Carrier signal, 240
Cascade form, 38–39, 45, 103
Cauchy theorem, 27
Causal restriction, 33–34
Cepstrum, 245
    complex, 245–247
    computation of, 251
    of rational z-transforms, 250–251
    recursive computation of, 247
    of vocal-tract impulse response, 259

Cesàro filter, 223
Chebyshev filters, 51, 56–61, 70–74
Chebyshev polynomials, 60
Chirp filtering, 213
Circuit speeds, 7
Circular convolution, 166–167, 205–206
Circular-shifting theorem, 167
Cn function (*see* Jacobian elliptic
    functions)
Comb filters, 79–84
Common arithmetic unit, 23–24
Comparison between analog and digital
    signal processing, 7–8
Compilers, 133, 152–158
Complex cepstrum (*see* Cepstrum)
Complex convolution theorem, 29–30
Compression:
    audio dynamic-range, 241, 253
    image dynamic-range, 254
Computers, small or medium sized, 9
Conformal mappings, 65–66, 70–78
Contrast enhancement of images, 253
Convergence:
    of two-sided z-transform, 34–35
    of z-transform, 26
Convolution:
    circular, 5
    direct, as filtering algorithm, 6
    discrete, as model of speech
        production, 243
    as filtering algorithm, 3
    as homomorphic system, 237
    linear, 5
    in noise calculations, 104
    as vector addition, 242–245
    z-transform of, 28
Cooley-Tukey algorithm (*see* Fast
    Fourier transform)
Correlation function, 121
Coupled form, 37–39, 45–46
Covariance function, estimation of,
    123–124

Deadband effect, 112–116
Decimation in frequency, 182–196
Decimation in time, 173–182
Deconvolution, 242–245
  of speech, 258–262
Delay:
  effect in cepstrum, 250
  due to invalid subroutine execution
    order, 149–152
  simulation of, 149
Difference equations:
  first-order, 21–24, 30–31
  mth-order, z-transform in solution of,
    39–44
  second-order, 31–33, 36–39
  time for, 7
Digital filtering, 2–6
  advantages of, 49
  approximation problem, 50
    Butterworth, 67, 72, 77
    Chebyshev, 73
    elliptic, 75, 86–90
    elliptic phase splitters, 90–92
  bandpass, 76–77
  bilinear transform, 51, 70–78, 86–87
  by fast convolution, 208, 217–231
  frequency-sampling, 51, 78–86
  impulse invariant, 51–58
  realizability of, 69–70, 93–97
  role of display in design of, 11–13
  synthesis problem, 50
  in system simulation, 131–133, 141–142
Digital networks:
  for first-order systems, 22–23
  for mth-order systems, 41–42, 44–46
  notation for, 22–23
  for second-order systems, 36–39
Digital resonator:
  execution time for, 7
  gain of, 55–56
  noise analysis of, 106–107
  zero of, 55–56
Direct form, 36
  inadvisability of implementation,
    103, 108
Discrete filters (see Digital filters)
Discrete Fourier transform, 3, 5–6, 251
  of artificially lengthened sequences,
    168–169
  of complex sinusoidal sequence, 164
  definition, 162

Discrete Fourier transform:
  inverse of, 165, 196
  as a lagged product, 213–217
  of odd or even sequences, 167–169
  of permuted sequences, 170–171
  of real or imaginary sequences,
    168–169
  relation to circular convolution,
    165–166
  relation to z-transform, 163
  of two real sequences, 197–199
Displays:
  computer controlled, 9–13
  of sampled waveforms, 140–141
  in system simulations, 154
Distortion of frequency scale, 71
Dithering to avoid deadband, 116
Dn function (see Jacobian elliptic
  functions)
Dynamic-range compression, image, 254

Echo removal, 253
Eigenvalues, 45
Elemental filters, 78–84
Elliptic filters, 58, 61–65, 75
  procedure for design of, 86–89
Elliptic functions (see Jacobian elliptic
  functions)
Elliptic phase splitting network, 90–92
Envelope, 240
Equal ripple filters, 60–66
Ergodic hypothesis, 126
Error criterion in generalized linear
  filtering, 239
Error effects (see Quantization effects)
Event detector (see Seismology)

Fading, 240
Fast Fourier transform, 2, 6, 173–201
  decimation in frequency, 182–196
    mixed radix case, 186–187
  decimation in time, 173–182
    mixed radix case, 182
Feedback in next-state simulation,
  151–152
Fejér filter, 223
Filter configurations, 5, 44–46
Filtering:
  of multiplied signals, 240–242
  optimal, 240
  (See also Digital filtering)

Flow chart, 132–134
Folding (*see* Aliasing)
Fourier transformation:
  as filtering technique, 3, 208
  properties, 159–162
  of a sequence, 161–162
Fourth-order filter, 112
Frequency response:
  of first-order network, 24–25
  geometric interpretation of, 42–43
Frequency sampling, 51, 78–86
  bank of filters, 84

Gaussian density, 145–146
Generalized linear filtering, 239–240
Generalized superposition, 235–238
Generators, 143–149
Geophysics (*see* Seismology)
Gibbs' phenomenon, 220, 226
Glottal pulse, 258
Goertzel algorithm, 171–172

High-speed convolution and correlation,
  2, 203–232
Hilbert transform, 247
Hilbert transformer, 222
Homomorphic filtering of convólved
  signals, 242–245
Homomorphic systems, 236–237

Illumination function, 253
Image processing, 241, 253
Impulse invariance (*see* Digital filtering)
Impulse response, 41
  duration of, 4, 6, 48–49, 82
  for frequency-sampling filters, 82
  of ideal low-pass filter, 35–36
In-place computation, 177–181, 185–186
Initial conditions, 43–44
Inputs to waveform processing
  simulations, 143
Interpolation by fast Fourier transform,
  199–200
Interpolation functions, 79
Inverse discrete Fourier transform,
  173–201
Inverse filtering, 243
Isolation, 39, 50

Jacobian elliptic functions, 61–65,
  87–92

Lagged products, 203–217
  for discrete Fourier transform
    computation, 213–217
  summation of, 203
Lerner filters, 52–54
Limit cycles, 148
Limiter (*see* Bandpass limiter)
Linear algebra, 236
Linear filtering, 239
Linear filters to measure noise, 122
Linear phase, 35, 82–83
  of frequency-sampling filters, 79
Linear systems, 235
Linear transformation, 236, 239
Logarithm, 3, 238
  of complex function, 241
Lossless resonators, 79–80

Man-machine communication, 8
Maximum phase, 249
Mean value of noise, 121
Memoryless systems, 237
Minimum phase, 247–249
Mirror-image polynomials, 93–97
Multipath, 242
Multiplication:
  scalar, 238
  as vector addition, 238–240

Networks, digital (*see* Digital networks)
Next-state simulation, 131–158
Noise measurements, 120–129, 133–135
Nonlinear systems, 233–235
Nonlinearities in system simulation, 139
Numerical filters (*see* Digital filters)
Nyquist rate, 137–140

Object program, 133
Object system, 133–136, 137–140
  elements for, 141–149
Output nodes, 134–136
Overlap-add method, 210
Overlap-save method, 209

Pairing in real convolutions, 211
Parallel form, 38–39, 45, 103
Parameter quantization, 39
Parseval's theorem, 196
Partial-fraction expansion, 44–45
Particular solution, 44
PATSI, 152–158
Periodic convolution (see Circular
   convolution)
Periodogram, 124
Peripheral devices attached to computers,
   8–9
Phase:
   computation of, 252
   minimum, 247–249
Phase splitting network, 90–92
Photographs, 3
Picture processing (see Image
   processing)
Poles, 30–33, 36, 37, 43–44
   cepstral effect of, 246
   as eigenvalues, 45
Primitive roots, 215
Principle of superposition, 235
Probability-density function, 243
Product, z-transform of, 29–30

Quantization effects, 3, 5, 20, 45,
   98–129
   deadband effect, 112–116
   in frequency-sampling filters, 83
   in parameters, 99–102
   from multiplications in digital
      filters, 108
   probability density of, 100–101
   in resonators, 106–107
   roundoff, dependence on network
      topology, 99
   sensitivity in narrowband low-
      pass filters, 102
   signal independence, 103
   in sin-cosine generators, 147–149
   in system simulation, 133
   sign-magnitude truncation, 99
   signal independence, 103
   in system simulation, 137
   truncation, 99

Radar, 1, 8, 9
Ramp generator, 143–144

Random-number generator, 137,
   144–146
Rayleigh density, 145–146
RC circuit, 22–25
Real part sufficiency, 248
Real time, 3, 7–8, 23–24, 136
Realizability, 69–70, 93–97
Recursion, as filtering algorithm, 3
Recursive filters, 48–49
Reflectance function, 253
Register lengths (see Quantization
   effects)
Residue theorem (see Cauchy theorem)
Reverberation, 243
RLC circuit, 32
Root-mean-square value of noise, 121
Roundoff (see Quantization effects)

Sampling, 20–21, 26
   display of waveforms in, 140–141
   effect of, in system simulation,
      137–140
   interval of, 3
   and quantization of parameters, 102
Sawtooth generator, 143–144
Scalar multiplication, 238
Sectioning, 208–210
   overlap-add method, 210
   overlap-save method, 209
Seismology, 1, 7–9
   event detector, 132–133
Serial form (see Cascade form)
Signal flow graph, 175
Simulation:
   next-state, 131–158
   of waveform processing systems,
      131–158
      inputs to, 143
      testing of, 143
Sin-cosine generator, 146–149
   noise considerations in, 147–149
Smith chart, 74–76
Sn function (see Jacobian elliptic
   functions)
Sonar, 1
Source program, 133, 153–158
Source system, 133, 136–140
Spatial filtering, 3
Special purpose hardware for filters, 49
Spectral estimation, 120–125

Spectrum analysis:
  discrete, 6–7
  of speech, 8
Speech, 1, 7, 243
  processing, homomorphic, 258–262
Square-wave generator, 144
Squared-magnitude function, 66–70
  realizability of, 69–70, 93–97
Stability, 4
  of first-order system, 28
  of second-order system, 33
Stationary process, 123
Step response of first-order system, 22
Subroutines in next-state simulation,
      141–158
  order of execution, 149–152
Superposition, 3, 235
  generalized, 235–238
Syllable segmentation, synchronous
      timing markers for, 9–11
Synthesis problem, 50
System function (*see* Transfer function)

Testing of waveform processing
      simulations, 143
Theta functions, 89
Time sharing of computers, 9
Time-varying coefficients, 49
Transfer function:
  for first-order systems, 24–25
  geometric interpretation of, 24–25,
      28–29, 42–43
  for mth-order systems, 41–43
  for second-order systems, 31–33
Twiddle factors, 194–195
Two-dimensional filtering, 3
Two-sided z-transform (*see* z-transform)
TX-2 computer, 8–9

Unit circle:
  in geometric interpretation of
      frequency response, 24–25, 42–43
  as integration path, 28, 30–31, 34–35
Unit-pulse response (*see* Impulse
      response)

Variance:
  of digital filter noise, 104
  nonconvergence for periodogram, 124
  of roundoff or truncation errors, 101
  of statistical measurements, 121–127
Vector space representation of signals,
      236
Vector space representation of systems,
      236
Vocal-tract impulse response, 243, 258
  cepstrum of, 259
Vowel generator, 127
Vowel spectrum, 259

Waveform processing, 239
  systems simulation, 131–158
    inputs to, 143
    testing of, 143
Weighting function, 226
Windows:
  Blackman, 228, 231
  Hamming, 228
  hanning, 228
  Kaiser's, 228
  rectangular, 226
  for spectral measurements, 125
  triangular, 226
Word-length requirements in digital
      filters, 105–107

z plane, 25–28, 32–33
z-transform:
  of a convolution, 28
  definition of, 26
  history of, 2
  inverse, 27–28
  of a product, 29–30
  in solution of difference equations,
      30–33, 39–44
  two-sided, 33–35, 244
Zeros, 36–37, 43–44
  cepstral effect of, 246
  in frequency-sampling filters, 83
  at infinity, 56–58